Modern Electrical Devices

Modern Electrical Devices is a woman-owned distribution company. We sell electrical supplies including terminals, cable ties, heat shrink tubing, fuses, and more.

Our goal is to provide outstanding service and products to our customers.

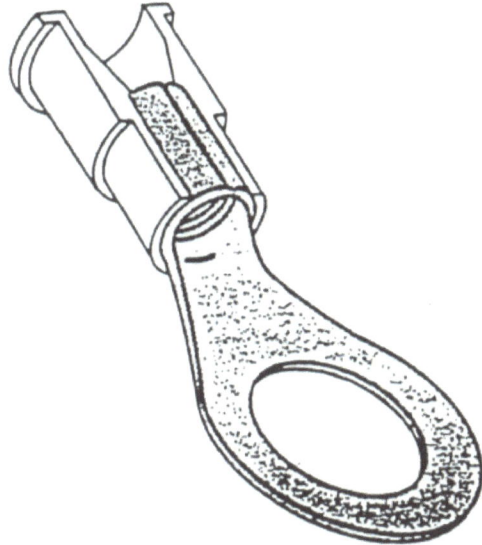

TERMINAL FEATURES

- One piece, burr free construction for economy, mechanical strength and maximum electrical conductivity
- Beveled terminal mouth to facilitate wire insertion on bare terminals
- Funnel barrel to facilitate wire insertion on insulated terminals
- Multiple "V" grooves for maximum holding power and minimum contact resistance
- Constructed from pure electrolyte copper, annealed to guard against breaking due to vibration and bending
- Uniform electro-tin plating, for maximum corrosion protection
- Full 1/4" long barrel provides extra length for fast, easy location when crimping
- Wire ranges clearly marked on all terminals
- UL & CSA listings available where applicable

TERMINAL DESCRIPTION

NON-INSULATED BUTTED SEAM: One piece, burr free, non brazed terminal for the economy minded user

NON-INSULATED BRAZED SEAM: Silver brazed barrel that can be crimped anywhere on the barrel surface. This patented process does not permit the seam to open under crimping pressure or operation stress

NON-INSULATED SEAMLESS: Manufactured from seamless electrolytic copper tubing. Can be crimped anywhere on the barrel because there is no seam to open

PLASTIC INSULATED: An assembly of a non-insulated part with a flared, rigid, polyvinyl chloride, color coded sleeve, securely attached and funneled for easy wire entry. Operating temperature range continuous duty from -40°F to 220° F (-40°C to 93°C). Intermittent duty to 300°F (149°C)

NYLON INSULATED: Consists of a non-insulated part securely attached to a flared, color coded, nylon sleeve. Operating temperature range- continuous duty from -40°F to 250°F (-40°C to 121°C). Intermittent duty to 350°F (177°C)

NYLON INSULATED WITH INSULATION GRIP: Constructed of a non-insulated part with a funnel entry, flared, seamless, metal insulation gripping sleeve securely attached and covered with a color coded flare nylon sleeve. This metal insulation gripping sleeve adds extra barrel strength and allows the wire to be sent in any direction without fraying the wires insulation or breaking the conductor. Operating temperature range from -40°F to 250°F (-40° to 121°C). Intermittent duty to 350°F (177°C)

TERMS & POLICIES:
Terms: Net 30 Days • Freight Policy • FOB: Waterloo, IL
Prepaid freight on orders $500.00 minimum • Returns subject to 15% restocking fee charge

Modern Electrical Devices

NYLON INSULATED WITH INSULATION GRIP

RING TONGUE TERMINALS

WIRE RANGE 26-24 (YELLOW)

STUD	PART NO.	STUD	PART NO.	STUD	PART NO.
0	NC24-0R	4	NC24-4R	8	NC24-8R
2	NC24-2R	6	NC24-6R	10	NC24-10R

WIRE RANGE 22-18 (RED)

STUD	PART NO.	STUD	PART NO.	STUD	PART NO.	STUD	PART NO.	STUD	PART NO.
2	NC18-2RSS	2	NC18-2RS	6	NC18-6R	1/4	NC18-14R	3/8	NC18-38R
4	NC18-4RSS	4	NC18-4RS	8	NC18-8R	5/16	NC18-56R		
6	NC18-6RSS	6	NC18-6RS	10	NC18-10R				

WIRE RANGE 16-14 (BLUE)

STUD	PART NO.	STUD	PART NO.	STUD	PART NO.	STUD	PART NO.
2	NC14-2RS	6	NC14-6R	1/4	NC14-14R	3/8	NC14-38R
4	NC14-4RS	8	NC14-8R	5/16	NC14-56R		
6	NC14-6RS	10	NC14-10R				

WIRE RANGE 12-10 (YELLOW)

STUD	PART NO.	STUD	PART NO.	STUD	PART NO.	STUD	PART NO.	STUD	PART NO.
4	NC10-4RS	6	NC10-6R	1/4	NC10-14R	3/8	NC10-38R	7/16	NC10-76R
6	NC10-6RS	8	NC10-8R	5/16	NC10-56R			1/2	NC10-12R
		10	NC10-10R						

Modern Electrical Devices

NYLON INSULATED WITH INSULATION GRIP

HEAVY DUTY RING TONGUE

WIRE RANGE 16-12 (YELLOW)

STUD	PART NO.	STUD	PART NO.	STUD	PART NO.	STUD	PART NO.	STUD	PART NO.
4	NC12-4RS	6	NC12-6R	1/4	NC12-14RS	1/4	NC12-14R	7/16	NC12-76R
6	NC12-6RS	8	NC12-8R	5/16	NC12-56RS	5/16	NC12-56R	1/2	NC12-12R
		10	NC12-10R			3/8	NC12-38R		

NYLON INSULATED BRAZED SEAM

RING TONGUE TERMINAL

WIRE SIZE #8 (RED)

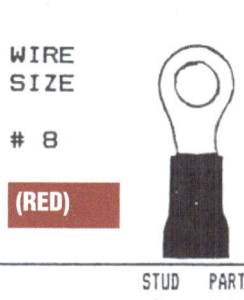     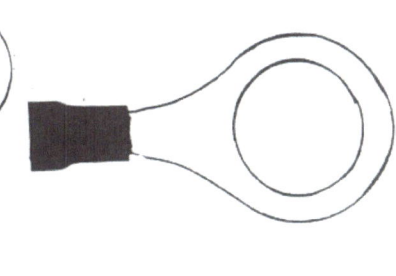

STUD	PART NO.	STUD	PART NO.	STUD	PART NO.	STUD	PART NO.
8	NW8-8R	10	NW8-10RW	7/16	NW8-76R	7/16	NW8-76RW
10	NW8-10R	1/4	NW8-14RW	1/2	NW8-12R	1/2	NW8-12RW
1/4	NW8-14R	5/16	NW8-56R			5/8	NW8-58R
		3/8	NW8-38R			3/4	NW8-34R

WIRE SIZE #6 (BLUE)

   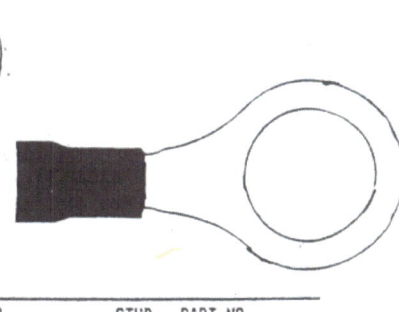

STUD	PART NO.	STUD	PART NO.	STUD	PART NO.	STUD	PART NO.
8	NW6-8R	10	NW6-10RW	7/16	NW6-76R	7/16	NW6-76RW
10	NW6-10R	1/4	NW6-14RW	1/2	NW6-12R	1/2	NW6-12RW
1/4	NW6-14R	5/16	NW6-56R			5/8	NW6-58R
		3/8	NW6-38R			3/4	NW6-34R

Modern Electrical Devices

NYLON INSULATED BRAZED SEAM

RING TONGUE TERMINAL

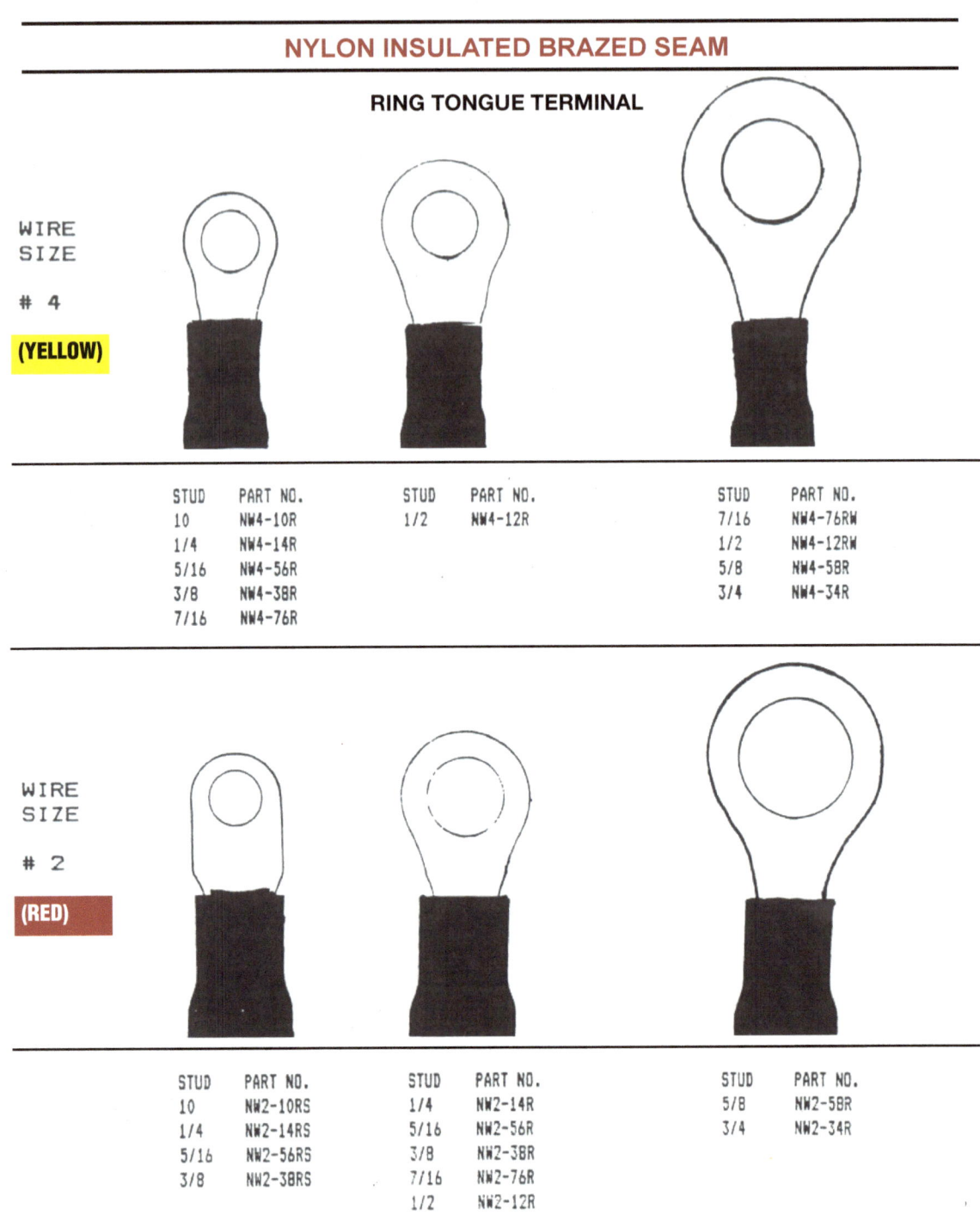

WIRE SIZE # 4 (YELLOW)

STUD	PART NO.	STUD	PART NO.	STUD	PART NO.
10	NW4-10R	1/2	NW4-12R	7/16	NW4-76RW
1/4	NW4-14R			1/2	NW4-12RW
5/16	NW4-56R			5/8	NW4-58R
3/8	NW4-38R			3/4	NW4-34R
7/16	NW4-76R				

WIRE SIZE # 2 (RED)

STUD	PART NO.	STUD	PART NO.	STUD	PART NO.
10	NW2-10RS	1/4	NW2-14R	5/8	NW2-58R
1/4	NW2-14RS	5/16	NW2-56R	3/4	NW2-34R
5/16	NW2-56RS	3/8	NW2-38R		
3/8	NW2-38RS	7/16	NW2-76R		
		1/2	NW2-12R		

NW NYLON INSULATED BRAZED SEAM #1/0 THROUGH #4/0 ALSO AVAILABLE

Modern Electrical Devices

NYLON INSULATED WITH INSULATION GRIP

WIRE RANGE 26-24 (YELLOW)

SPADE	SPADE	SPADE
STUD 0 — NC24-0S	STUD 4 — NC24-4S	STUD 6 — NC24-6S

WIRE RANGE 22-18 (RED)

SPADE	BLOCK SPADE	FLANGED BLOCK SPADE	HOOKS
STUD / PART NO.	STUD / PART NO.	STUD / PART NO.	STUD / PART NO.
6 NC18-6S	6 NC18-6SN	6 NC18-6F	6 NC18-6H
8 NC18-8S	8 NC18-8SN	8 NC18-8F	8 NC18-8H
10 NC18-10S	10 NC18-10SN	10 NC18-10F	10 NC18-10H

WIRE RANGE 16-14 (BLUE)

SPADE	BLOCK SPADE	FLANGED BLOCK SPADE	HOOKS
STUD / PART NO.	STUD / PART NO.	STUD / PART NO.	STUD / PART NO.
6 NC14-6S	6 NC14-6SN	6 NC14-6F	6 NC14-6H
8 NC14-8S	8 NC14-8SN	8 NC14-8F	8 NC14-8H
10 NC14-10S	10 NC14-10SN	10 NC14-10F	10 NC14-10H

WIRE RANGE 12-10 (YELLOW)

SPADE	BLOCK SPADE	FLANGED BLOCK SPADE	HOOKS
STUD / PART NO.	STUD / PART NO.	STUD / PART NO.	STUD / PART NO.
6 NC10-6S	6 NC10-6SN	6 NC10-6F	6 NC10-6H
8 NC10-8S	8 NC10-8SN	8 NC10-8F	8 NC10-8H
10 NC10-10S	10 NC10-10SN	10 NC10-10F	10 NC10-10H
1/4 NC10-14S			1/4 NC10-14H
5/16 NC10-56S			5/16 NC10-56H

ModernElectricalDevices.com   Office: 314-443-2943

Modern Electrical Devices

NYLON INSULATED WITH INSULATION GRIP

HEAVY DUTY SPADE

WIRE RANGE 16-12 **(YELLOW)**

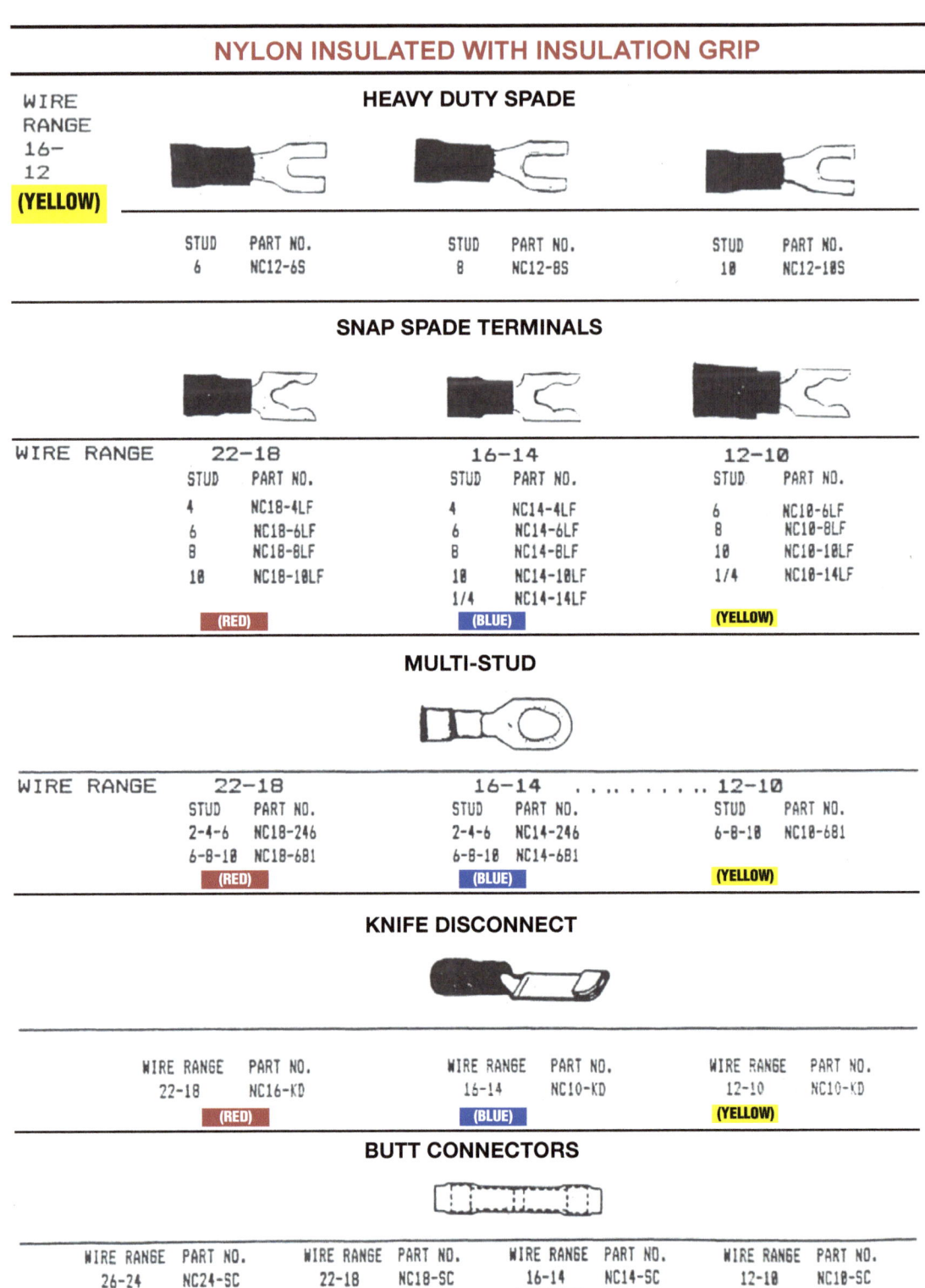

STUD	PART NO.	STUD	PART NO.	STUD	PART NO.
6	NC12-6S	8	NC12-8S	10	NC12-10S

SNAP SPADE TERMINALS

WIRE RANGE	22-18		16-14		12-10	
	STUD	PART NO.	STUD	PART NO.	STUD	PART NO.
	4	NC18-4LF	4	NC14-4LF	6	NC10-6LF
	6	NC18-6LF	6	NC14-6LF	8	NC10-8LF
	8	NC18-8LF	8	NC14-8LF	10	NC10-10LF
	10	NC18-10LF	10	NC14-10LF	1/4	NC10-14LF
			1/4	NC14-14LF		
	(RED)		**(BLUE)**		**(YELLOW)**	

MULTI-STUD

WIRE RANGE	22-18		16-14		12-10	
	STUD	PART NO.	STUD	PART NO.	STUD	PART NO.
	2-4-6	NC18-246	2-4-6	NC14-246	6-8-10	NC10-681
	6-8-10	NC18-681	6-8-10	NC14-681		
	(RED)		**(BLUE)**		**(YELLOW)**	

KNIFE DISCONNECT

WIRE RANGE	PART NO.	WIRE RANGE	PART NO.	WIRE RANGE	PART NO.
22-18	NC16-KD	16-14	NC10-KD	12-10	NC10-KD
(RED)		**(BLUE)**		**(YELLOW)**	

BUTT CONNECTORS

WIRE RANGE	PART NO.	WIRE RANGE	PART NO.	WIRE RANGE	PART NO.	WIRE RANGE	PART NO.
26-24	NC24-SC	22-18	NC18-SC	16-14	NC14-SC	12-10	NC10-SC
(YELLOW)		**(RED)**		**(BLUE)**		**(YELLOW)**	

Modern Electrical Devices

NYLON INSULATED WITH INSULATION GRIP

WINDOW CONNECTORS

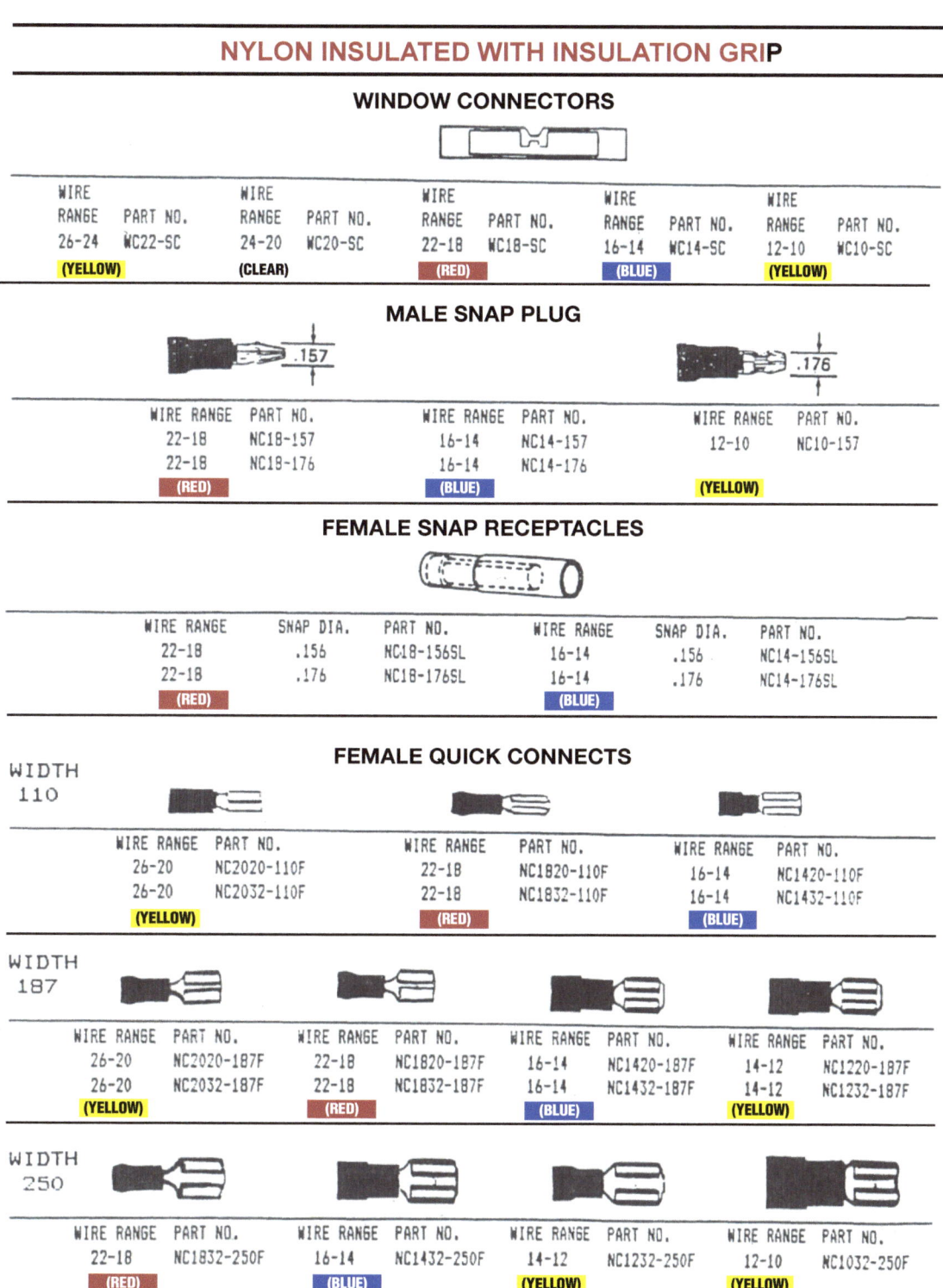

WIRE RANGE	PART NO.	WIRE RANGE	PART NO.	WIRE RANGE	PART NO.	WIRE RANGE	PART NO.	WIRE RANGE	PART NO.
26-24	WC22-SC	24-20	WC20-SC	22-18	WC18-SC	16-14	WC14-SC	12-10	WC10-SC
(YELLOW)		(CLEAR)		(RED)		(BLUE)		(YELLOW)	

MALE SNAP PLUG

WIRE RANGE	PART NO.	WIRE RANGE	PART NO.	WIRE RANGE	PART NO.
22-18	NC18-157	16-14	NC14-157	12-10	NC10-157
22-18	NC18-176	16-14	NC14-176		
(RED)		(BLUE)		(YELLOW)	

FEMALE SNAP RECEPTACLES

WIRE RANGE	SNAP DIA.	PART NO.	WIRE RANGE	SNAP DIA.	PART NO.
22-18	.156	NC18-156SL	16-14	.156	NC14-156SL
22-18	.176	NC18-176SL	16-14	.176	NC14-176SL
(RED)			(BLUE)		

FEMALE QUICK CONNECTS

WIDTH 110

WIRE RANGE	PART NO.	WIRE RANGE	PART NO.	WIRE RANGE	PART NO.
26-20	NC2020-110F	22-18	NC1820-110F	16-14	NC1420-110F
26-20	NC2032-110F	22-18	NC1832-110F	16-14	NC1432-110F
(YELLOW)		(RED)		(BLUE)	

WIDTH 187

WIRE RANGE	PART NO.	WIRE RANGE	PART NO.	WIRE RANGE	PART NO.	WIRE RANGE	PART NO.
26-20	NC2020-187F	22-18	NC1820-187F	16-14	NC1420-187F	14-12	NC1220-187F
26-20	NC2032-187F	22-18	NC1832-187F	16-14	NC1432-187F	14-12	NC1232-187F
(YELLOW)		(RED)		(BLUE)		(YELLOW)	

WIDTH 250

WIRE RANGE	PART NO.	WIRE RANGE	PART NO.	WIRE RANGE	PART NO.	WIRE RANGE	PART NO.
22-18	NC1832-250F	16-14	NC1432-250F	14-12	NC1232-250F	12-10	NC1032-250F
(RED)		(BLUE)		(YELLOW)		(YELLOW)	

Modern Electrical Devices

NYLON INSULATED WITH INSULATION GRIP

MALE QUICK CONNECTS

WIDTH 187

WIRE RANGE	PART NO.	WIRE RANGE	PART NO.
22-18	NC1820-187M	16-14	NC1420-187M
22-18	NC1832-187M	16-14	NC1432-187M
(RED)		(BLUE)	

WIDTH 250

WIRE RANGE	PART NO.	WIRE RANGE	PART NO.	WIRE RANGE	PART NO.
22-18	NC1832-250M	16-14	NC1432-250M	12-10	NC1032-250M
(RED)		(BLUE)		(YELLOW)	

MULTI STACK (PIGGYBACKS)

WIDTH 250

WIRE RANGE	PART NO.	WIRE RANGE	PART NO.
22-18	NC1832-PB	16-14	NC1432-PB
(RED)		(BLUE)	

FULLY INSULATED FEMALE COUPLER

WIDTH 187

WIRE RANGE	PART NO.	WIRE RANGE	PART NO.
22-18	NCF1820-187F	16-14	NCF1420-187F
(RED)		(BLUE)	

WIDTH 250

WIRE RANGE	PART NO.	WIRE RANGE	PART NO.	WIRE RANGE	PART NO.
22-18	NCF1832-250F	16-14	NCF1432-250F	12-10	NCF1032-250F
(RED)		(BLUE)		(YELLOW)	

FULLY INSULATED MALE COUPLER

WIDTH 187

WIRE RANGE	PART NO.	WIRE RANGE	PART NO.
22-18	NCF1820-187M	16-14	NCF1420-187M
(RED)		(BLUE)	

WIDTH 250

WIRE RANGE	PART NO.	WIRE RANGE	PART NO.	WIRE RANGE	PART NO.
22-18	NCF1832-250M	16-14	NCF1432-250M	12-10	NCF1032-250M
(RED)		(BLUE)		(YELLOW)	

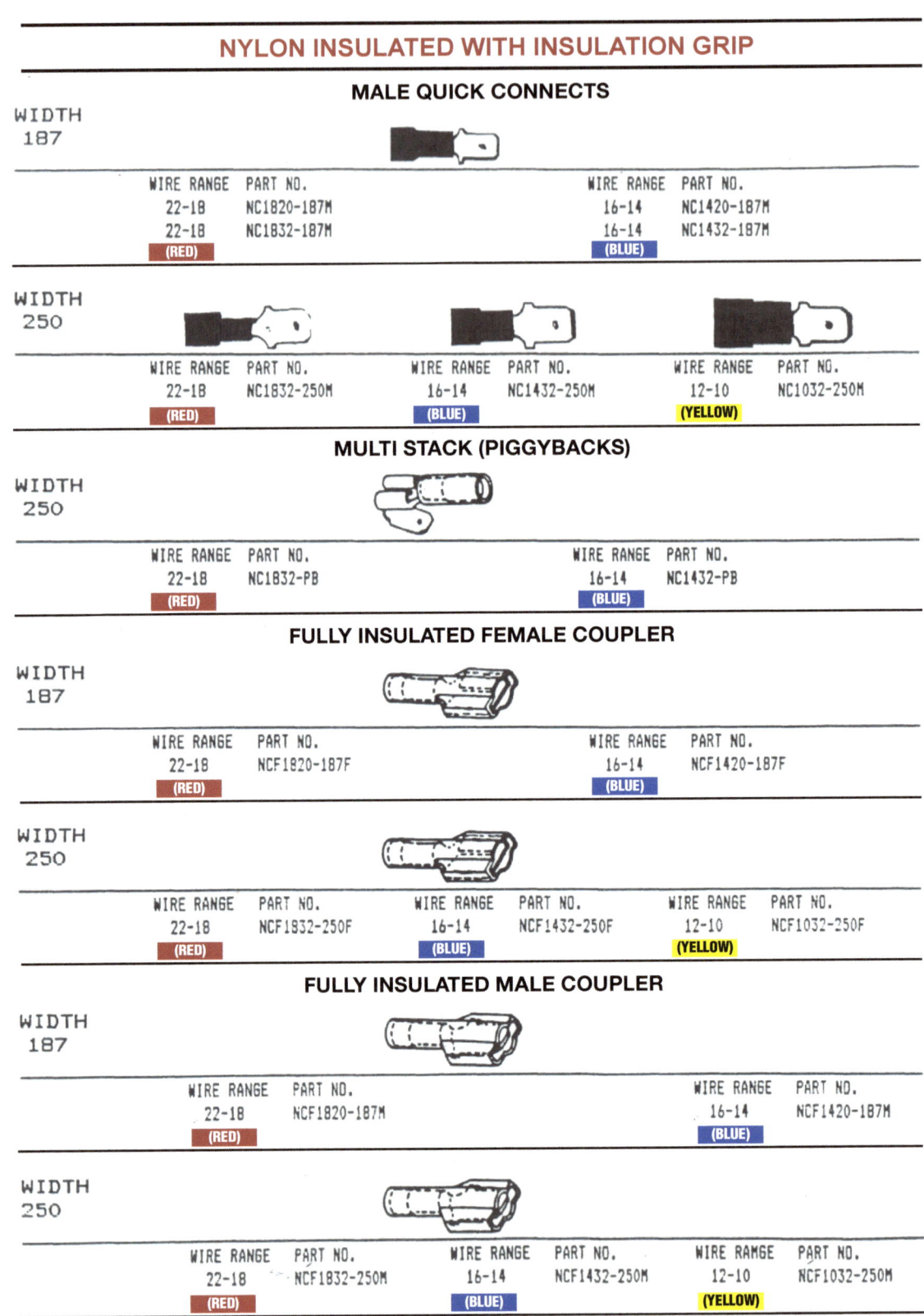

Modern Electrical Devices

NYLON INSULATED WITH INSULATION GRIP

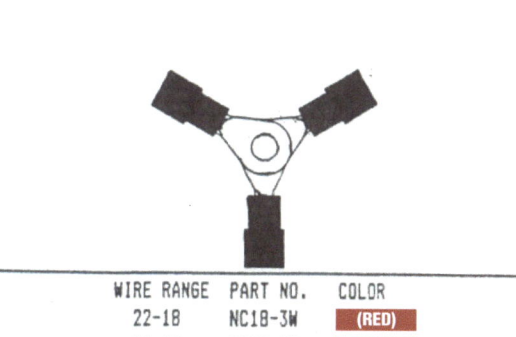

3-WAY CONNECTORS . . . . 4-WAY CONNECTORS

WIRE RANGE	PART NO.	COLOR	WIRE RANGE	PART NO.	COLOR
22-18	NC18-3W	(RED)	22-18	NC18-4W	(RED)
16-14	NC18-3W	(BLUE)	16-14	NC14-4W	(BLUE)
12-10	NC10-3W	(YELLOW)	12-10	NC10-4W	(YELLOW)

NYLON INSULATED

FULLY INSULATED FEMALE COUPLER

WIDTH 187

WIRE RANGE	PART NO.	WIRE RANGE	PART NO.
22-18 (RED)	NF1820-187F	16-14 (BLUE)	NF1420-187F

WIDTH 250

WIRE RANGE	PART NO.	WIRE RANGE	PART NO.	WIRE RANGE	PART NO.
22-18 (RED)	NF1832-250F	16-14 (BLUE)	NF1432-250F	12-10 (YELLOW)	NF1032-250F

FULLY INSULATED MALE COUPLER

WIDTH 187

WIRE RANGE	PART NO.	WIRE RANGE	PART NO.
22-18 (RED)	NF1820-187M	16-14 (BLUE)	NF1420-187M

WIDTH 250

WIRE RANGE	PART NO.	WIRE RANGE	PART NO.	WIRE RANGE	PART NO.
22-18 (RED)	NF1832-250M	16-14 (BLUE)	NF1432-250M	12-10 (YELLOW)	NF1032-250M

Modern Electrical Devices

NYLON INSULATED

FULLY INSULATED FEMALE FLAG DISCONNECT

WIDTH 187

WIRE RANGE	PART NO.		WIRE RANGE	PART NO.
22-18	NF1820-187FL		16-14	NF1420-187FL
(RED)			(BLUE)	

WIDTH 250

WIRE RANGE	PART NO.		WIRE RANGE	PART NO.
22-18	NF1832-250FL		16-14	NF1432-250FL
(RED)			(BLUE)	

FEMALE SNAP RECEPTACLES

WIRE RANGE	SNAP DIA.	PART NO.	WIRE RANGE	SNAP DIA.	PART NO.
22-18	.157	N18-157SL	16-14	.157	N14-157SL
22-18	.176	N18-176SL	16-14	.176	N14-176SL
(RED)			(BLUE)		

CLOSED-END CONNECTORS

WIRE RANGE	PART NO.	WIRE RANGE	PART NO.	WIRE RANGE	PART NO.
22-14	N2214-CE	16-10	N1610-CE	18-8	N128-CE

BUTT CONNECTORS

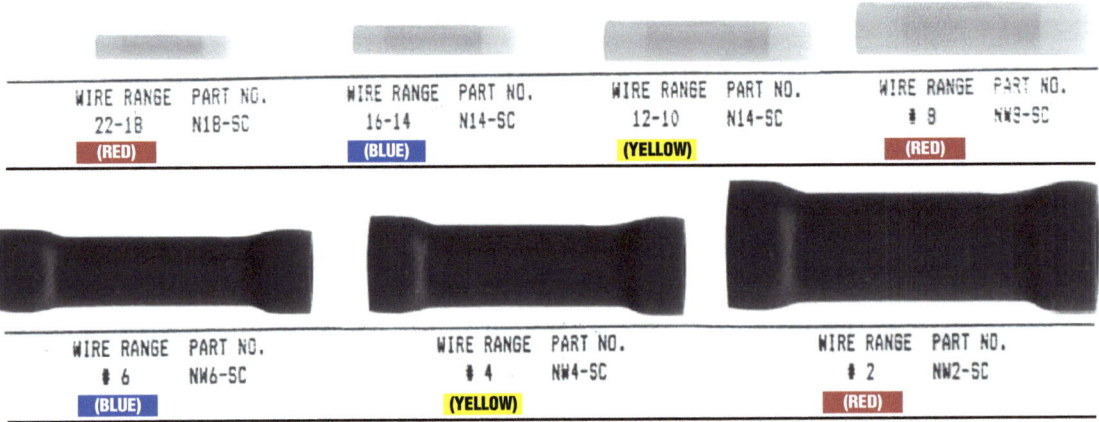

WIRE RANGE	PART NO.	WIRE RANGE	PART NO.	WIRE RANGE	PART NO.	WIRE RANGE	PART NO.
22-18	N18-SC	16-14	N14-SC	12-10	N14-SC	# 8	NW8-SC
(RED)		(BLUE)		(YELLOW)		(RED)	

WIRE RANGE	PART NO.	WIRE RANGE	PART NO.	WIRE RANGE	PART NO.
# 6	NW6-SC	# 4	NW4-SC	# 2	NW2-SC
(BLUE)		(YELLOW)		(RED)	

Modern Electrical Devices

BUTT CONNECTORS (Continued)

WIRE RANGE   PART NO.
1/0   NW1/0-SC
(BLUE)

WIRE RANGE   PART NO.
2/0   NW2/0-SC
(YELLOW)

WIRE RANGE   PART NO.
3/0   NW3/0-SC
(RED)

WIRE RANGE   PART NO.
4/0   NW4/0-SC
(BLUE)

PARALLEL CONNECTORS

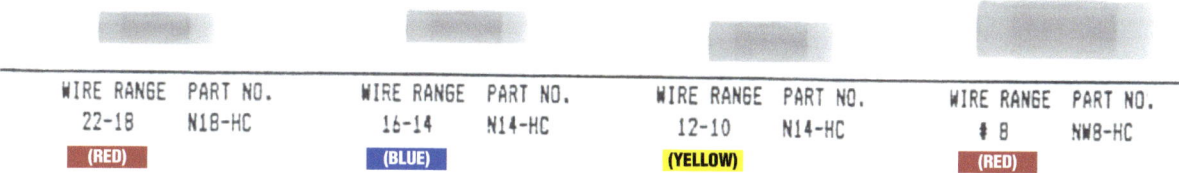

WIRE RANGE	PART NO.	WIRE RANGE	PART NO.	WIRE RANGE	PART NO.	WIRE RANGE	PART NO.
22-18	N18-HC	16-14	N14-HC	12-10	N14-HC	# 8	NW8-HC
(RED)		**(BLUE)**		**(YELLOW)**		**(RED)**	

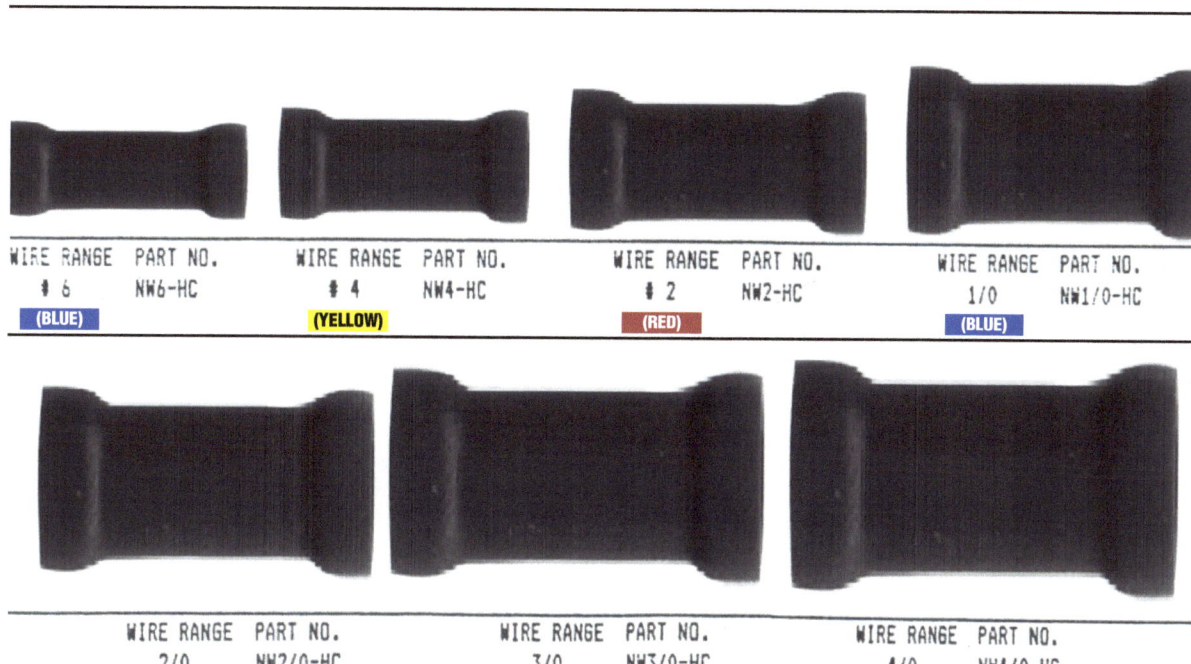

WIRE RANGE	PART NO.	WIRE RANGE	PART NO.	WIRE RANGE	PART NO.	WIRE RANGE	PART NO.
# 6	NW6-HC	# 4	NW4-HC	# 2	NW2-HC	1/0	NW1/0-HC
(BLUE)		**(YELLOW)**		**(RED)**		**(BLUE)**	

WIRE RANGE	PART NO.	WIRE RANGE	PART NO.	WIRE RANGE	PART NO.
2/0	NW2/0-HC	3/0	NW3/0-HC	4/0	NW4/0-HC
(YELLOW)		**(RED)**		**(BLUE)**	

ModernElectricalDevices.com   Office: 314-443-2943

Modern Electrical Devices

PLASTIC INSULATED FUNNEL ENTRY

RING TONGUE TERMINALS

WIRE RANGE 22-18 (RED)

STUD	PART NO.	STUD	PART NO.	STUD	PART NO.	STUD	PART NO.	STUD	PART NO.
2	C18-2RSS	2	C18-2RS	6	C18-6R	1/4	C18-14R	3/8	C18-38R
4	C18-4RSS	4	C18-4RS	8	C18-8R	5/16	C18-56R		
6	C18-6RSS	6	C18-6RS	10	C18-10R				

WIRE RANGE 16-14 (BLUE)

STUD	PART NO.	STUD	PART NO.	STUD	PART NO.	STUD	PART NO.
2	C14-2RS	6	C14-6R	1/4	C14-14R	3/8	C14-38R
4	C14-4RS	8	C14-8R	5/16	C14-56R		
6	C14-6RS	10	C14-10R				

WIRE RANGE 12-10 (YELLOW)

STUD	PART NO.	STUD	PART NO.	STUD	PART NO.	STUD	PART NO.	STUD	PART NO.
4	C10-4RS	6	C10-6R	1/4	C10-14R	3/8	C10-38R	7/16	C10-76R
6	C10-6RS	8	C10-8R	5/16	C10-56R			1/2	C10-12R
		10	C10-10R						

HEAVY DUTY RING TONGUE

WIRE RANGE 16-12 (YELLOW)

STUD	PART NO.	STUD	PART NO.	STUD	PART NO.	STUD	PART NO.	STUD	PART NO.
4	C12-4RS	6	C12-6R	1/4	C12-14RS	1/4	C12-14R	7/16	C12-76R
6	C12-6RS	8	C12-8R	5/16	C12-56RS	5/16	C12-56R	1/2	C12-12R
		10	C12-10R			3/8	C12-38R		

Modern Electrical Devices

PLASTIC INSULATED BRAZED SEAM

RING TONGUE TERMINALS

WIRE SIZE # 8 (RED)

STUD	PART NO.	STUD	PART NO.	STUD	PART NO.	STUD	PART NO.
8	CW8-8R	10	CW8-10RW	7/16	CW8-76R	7/16	CW8-76RW
10	CW8-10R	1/4	CW8-14RW	1/2	CW8-12R	1/2	CW8-12RW
1/4	CW8-14R	5/16	CW8-56R			5/8	CW8-58R
		3/8	CW8-38R			3/4	CW8-34R

WIRE SIZE # 6 (BLUE)

STUD	PART NO.	STUD	PART NO.	STUD	PART NO.	STUD	PART NO.
8	CW6-8R	10	CW6-10RW	7/16	CW6-76R	7/16	CW6-76RW
10	CW6-10R	1/4	CW6-14RW	1/2	CW6-12R	1/2	CW6-12RW
1/4	CW6-14R	5/16	CW6-56R			5/8	CW6-58R
		3/8	CW6-38R			3/4	CW6-34R

WIRE SIZE # 4 (YELLOW)

STUD	PART NO.	STUD	PART NO.	STUD	PART NO.
10	CW4-10R	1/2	CW4-12R	7/16	CW4-76RW
1/4	CW4-14R			1/2	CW4-12RW
5/16	CW4-56R			5/8	CW4-58R
3/8	CW4-38R			3/4	CW4-34R
7/16	CW4-76R				

CW PLASTIC INSULATED TERMINALS #2 THROUGH #4/0 ALSO AVAILABLE

Modern Electrical Devices

PLASTIC INSULATED FUNNEL ENTRY

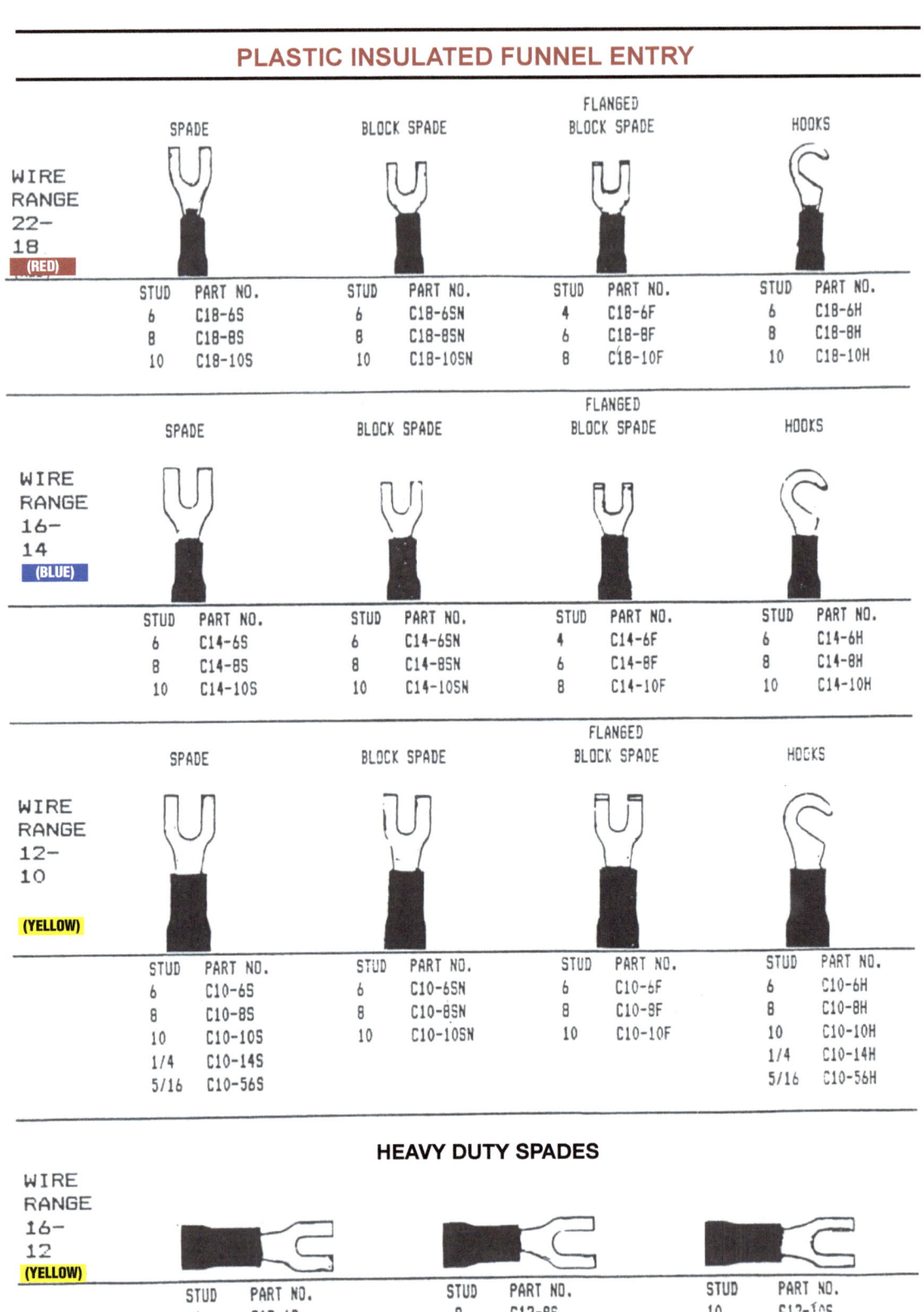

WIRE RANGE 22-18 (RED)

SPADE		BLOCK SPADE		FLANGED BLOCK SPADE		HOOKS	
STUD	PART NO.	STUD	PART NO.	STUD	PART NO.	STUD	PART NO.
6	C18-6S	6	C18-6SN	4	C18-6F	6	C18-6H
8	C18-8S	8	C18-8SN	6	C18-8F	8	C18-8H
10	C18-10S	10	C18-10SN	8	C18-10F	10	C18-10H

WIRE RANGE 16-14 (BLUE)

SPADE		BLOCK SPADE		FLANGED BLOCK SPADE		HOOKS	
STUD	PART NO.	STUD	PART NO.	STUD	PART NO.	STUD	PART NO.
6	C14-6S	6	C14-6SN	4	C14-6F	6	C14-6H
8	C14-8S	8	C14-8SN	6	C14-8F	8	C14-8H
10	C14-10S	10	C14-10SN	8	C14-10F	10	C14-10H

WIRE RANGE 12-10 (YELLOW)

SPADE		BLOCK SPADE		FLANGED BLOCK SPADE		HOOKS	
STUD	PART NO.	STUD	PART NO.	STUD	PART NO.	STUD	PART NO.
6	C10-6S	6	C10-6SN	6	C10-6F	6	C10-6H
8	C10-8S	8	C10-8SN	8	C10-8F	8	C10-8H
10	C10-10S	10	C10-10SN	10	C10-10F	10	C10-10H
1/4	C10-14S					1/4	C10-14H
5/16	C10-56S					5/16	C10-56H

HEAVY DUTY SPADES

WIRE RANGE 16-12 (YELLOW)

STUD	PART NO.	STUD	PART NO.	STUD	PART NO.
6	C12-6S	8	C12-8S	10	C12-10S

Modern Electrical Devices

PLASTIC INSULATED FUNNEL ENTRY

MULTI-STUD

WIRE RANGE 22-18	16-14	.12-10
STUD PART NO.	STUD PART NO.	STUD PART NO.
2-4-6 C18-246	2-4-6 C14-246	6-8-10 C10-681
6-8-10 C18-681	6-8-10 C14-681	
(RED)	(BLUE)	(YELLOW)

SNAP SPADE TERMINALS

WIRE RANGE 22-18	 16-14	 12-10
STUD PART NO.	STUD PART NO.	STUD PART NO.
4 C18-4LF	4 C14-4LF	6 C10-6LF
6 C18-6LF	6 C14-6LF	8 C10-8LF
8 C18-8LF	8 C14-8LF	10 C10-10LF
10 C18-10LF	10 C14-10LF	1/4 C10-14LF
	14 C14-14LF	
(RED)	(BLUE)	(YELLOW)

MALE SNAP PLUB

WIRE RANGE PART NO.	WIRE RANGE PART NO.	WIRE RANGE PART NO.
22-18 C18-157	16-14 C14-157	12-10 C10-157
22-18 C18-176	16-14 C14-176	
(RED)	(BLUE)	(YELLOW)

MULTI STACK (PIGGYBACKS)

WIRE RANGE PART NO.	WIRE RANGE PART NO.
22-18 C1832-PB	16-14 C1432-PB
(RED)	(BLUE)

Modern Electrical Devices

PLASTIC INSULATED FUNNEL ENTRY

FEMALE QUICK CONNECTS

WIDTH 110

WIRE RANGE	PART NO.	WIRE RANGE	PART NO.	WIRE RANGE	PART NO.
26-20	C2020-110F	22-18	C1820-110F	16-14	C1420-110F
26-20	C2032-110F	22-18	C1832-110F	16-14	C1432-110F
(YELLOW)		(RED)		(BLUE)	

WIDTH 187

WIRE RANGE	PART NO.	WIRE RANGE	PART NO.	WIRE RANGE	PART NO.	WIRE RANGE	PART NO.
26-20	C2020-187F	22-18	C1820-187F	16-14	C1420-187F	14-12	C1220-187F
26-20	C2032-187F	22-18	C1832-187F	16-14	C1432-187F	14-12	C1232-187F
(YELLOW)		(RED)		(BLUE)		(YELLOW)	

WIDTH 250

WIRE RANGE	PART NO.	WIRE RANGE	PART NO.	WIRE RANGE	PART NO.	WIRE RANGE	PART NO.
22-18	C1832-250F	16-14	C1432-250F	14-12	C1232-250F	12-10	C1032-250F
(RED)		(BLUE)		(YELLOW)		(YELLOW)	

MALE QUICK CONNECTS

WIDTH 187

WIRE RANGE	PART NO.	WIRE RANGE	PART NO.
22-18	C1820-187M	16-14	C1420-187M
22-18	C1832-187M	16-14	C1432-187M
(RED)		(BLUE)	

WIDTH 250

WIRE RANGE	PART NO.	WIRE RANGE	PART NO.	WIRE RANGE	PART NO.
22-18	C1832-250M	16-14	C1432-250M	12-10	C1032-250M
(RED)		(BLUE)		(YELLOW)	

Modern Electrical Devices

PLASTIC INSULATED FUNNEL ENTRY

3-WAY CONNECTORS . . . . 4-WAY CONNECTORS

WIRE RANGE	PART NO.	COLOR		WIRE RANGE	PART NO.	COLOR
22-18	C18-3W	(RED)		22-18	C18-4W	(RED)
16-14	C18-3W	(BLUE)		16-14	C14-4W	(BLUE)
12-10	C10-3W	(YELLOW)		12-10	C10-4W	(YELLOW)

PLASTIC INSULATED

BUTT CONNECTORS

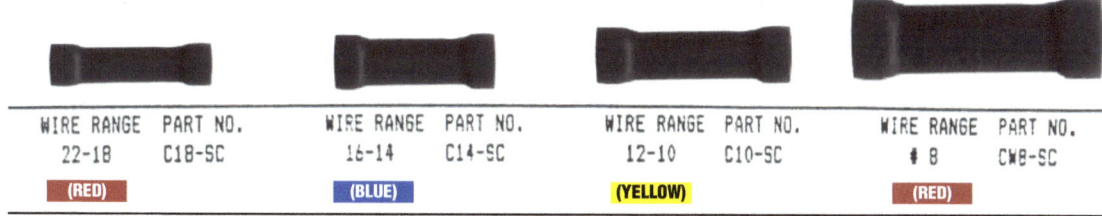

WIRE RANGE	PART NO.		WIRE RANGE	PART NO.		WIRE RANGE	PART NO.		WIRE RANGE	PART NO.
22-18	C18-SC		16-14	C14-SC		12-10	C10-SC		# 8	CW8-SC
(RED)			(BLUE)			(YELLOW)			(RED)	

WIRE RANGE	PART NO.		WIRE RANGE	PART NO.		WIRE RANGE	PART NO.
# 6	CW6-SC		# 4	CW4-SC		# 2	CW2-SC
(BLUE)			(YELLOW)			(RED)	

Modern Electrical Devices

PLASTIC INSULATED

BUTT CONNECTORS

WIRE RANGE   PART NO.
1/0   CW1/0-SC
(BLUE)

WIRE RANGE   PART NO.
2/0   CW2/0-SC
(YELLOW)

WIRE RANGE   PART NO.
3/0   CW3/0-SC
(RED)

WIRE RANGE   PART NO.
4/0   CW4/0-SC
(BLUE)

PARALLEL CONNECTORS

WIRE RANGE   PART NO.
22-18   C18-HC
(RED)

WIRE RANGE   PART NO.
16-14   C14-HC
(BLUE)

WIRE RANGE   PART NO.
12-10   C10-HC
(YELLOW)

WIRE RANGE   PART NO.
8   CW8-HC
(RED)

WIRE RANGE   PART NO.
6   CW6-HC
(BLUE)

WIRE RANGE   PART NO.
4   CW4-HC
(YELLOW)

WIRE RANGE   PART NO.
2   CW2-HC
(RED)

WIRE RANGE   PART NO.
1/0   CW1/0-HC
(BLUE)

WIRE RANGE   PART NO.
2/0   CW2/0-HC
(YELLOW)

WIRE RANGE   PART NO.
3/0   CW3/0-HC
(RED)

WIRE RANGE   PART NO.
4/0   CW4/0-HC
(BLUE)

Modern Electrical Devices

NON-INSULATED WITH BRAZED SEAM

RING TONGUE TERMINALS

WIRE RANGE 22-18

STUD	PART NO.	STUD	PART NO.	STUD	PART NO.	STUD	PART NO.	STUD	PART NO.
2	W18-2RSS	2	W18-2RS	6	W18-6R	1/4	W18-14R	3/8	W18-38R
4	W18-4RSS	4	W18-4RS	8	W18-8R	5/16	W18-56R		
6	W18-6RSS	6	W18-6RS	10	W18-10R				

WIRE RANGE 16-14

STUD	PART NO.	STUD	PART NO.	STUD	PART NO.	STUD	PART NO.
2	W14-2RS	6	W14-6R	1/4	W14-14R	3/8	W14-38R
4	W14-4RS	8	W14-8R	5/16	W14-56R		
6	W14-6RS	10	W14-10R				

WIRE RANGE 12-10

STUD	PART NO.	STUD	PART NO.	STUD	PART NO.	STUD	PART NO.	STUD	PART NO.
4	W10-4RS	6	W10-6R	1/4	W10-14R	3/8	W10-38R	7/16	W10-76R
6	W10-6RS	8	W10-8R	5/16	W10-56R			1/2	W10-12R
		10	W10-10R						

HEAVY DUTY RING TONGUE

WIRE RANGE 16-12

STUD	PART NO.	STUD	PART NO.	STUD	PART NO.	STUD	PART NO.	STUD	PART NO.
4	W12-4RS	6	W12-6R	1/4	W12-14R	1/4	W12-14R	7/16	W12-76R
6	W12-6RS	8	W12-8R	5/16	W12-56R	5/16	W12-56R	1/2	W12-12R
		10	W12-10R			3/8	W12-38R		

Modern Electrical Devices

NON-INSULATED WITH BRAZED SEAM

RING TONGUE TERMINALS

WIRE SIZE # 8

STUD	PART NO.	STUD	PART NO.	STUD	PART NO.	STUD	PART NO.
8	W8-8R	10	W8-10RW	7/16	W8-76R	7/16	W8-76RW
10	W8-10R	1/4	W8-14RW	1/2	W8-12R	1/2	W8-12RW
1/4	W8-14R	5/16	W8-56R			5/8	W8-58R
		3/8	W8-38R			3/4	W8-34R

WIRE SIZE # 6

STUD	PART NO.	STUD	PART NO.	STUD	PART NO.	STUD	PART NO.
8	W6-8R	10	W6-10RW	7/16	W6-76R	7/16	W6-76RW
10	W6-10R	1/4	W6-14RW	1/2	W6-12R	1/2	W6-12RW
1/4	W6-14R	5/16	W6-56R			5/8	W6-58R
		3/8	W6-38R			3/4	W6-34R

WIRE SIZE # 4

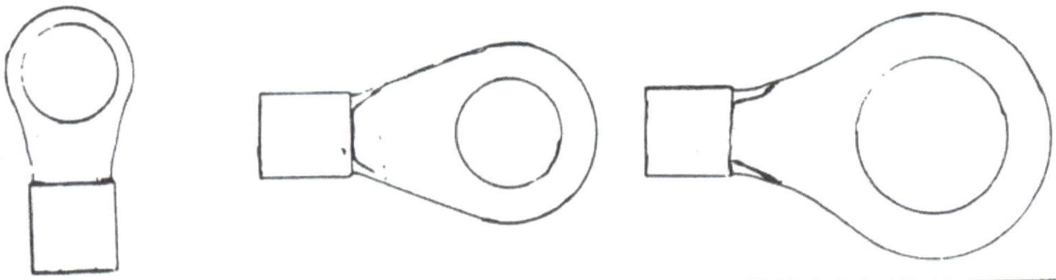

STUD	PART NO.	STUD	PART NO.	STUD	PART NO.
10	W4-10R	1/2	W4-12R	7/16	W4-76RW
1/4	W4-14R			1/2	W4-12RW
5/16	W4-56R			5/8	W4-58R
3/8	W4-38R			3/4	W4-34R
7/16	W4-76R				

ModernElectricalDevices.com   Office: 314-443-2943

Modern Electrical Devices

NON-INSULATED WITH BRAZED SEAM

RING TONGUE TERMINALS

WIRE SIZE # 2

STUD	PART NO.	STUD	PART NO.	STUD	PART NO.
10	W2-10RS	1/4	W2-14R	5/8	W2-58R
1/4	W2-14RS	5/16	W2-56R	3/4	W2-34R
5/16	W2-56RS	3/8	W2-38R		
3/8	W2-38RS	7/16	W2-76R		

WIRE SIZE 1/0

STUD	PART NO.	STUD	PART NO.
1/4	W1/0-14R	5/8	W1/0-58R
5/16	W1/0-56R	3/4	W1/0-34R
3/8	W1/0-38R		
7/16	W1/0-76R		

WIRE SIZE 2/0

STUD	PART NO.	STUD	PART NO.
1/4	W2/0-14R	5/8	W2/0-58R
5/16	W2/0-56R	3/4	W2/0-34R
3/8	W2/0-38R		
7/16	W2/0-76R		
1/2	W2/0-12R		

Modern Electrical Devices

NON-INSULATED WITH BRAZED SEAM

RING TONGUE TERMINALS

WIRE SIZE 3/0

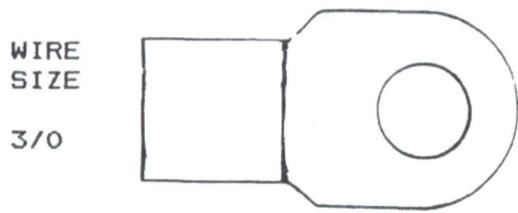 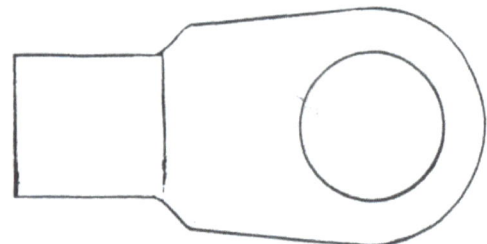

STUD	PART NO.		STUD	PART NO.
1/4	W3/0-14R		1/2	W3/0-12RW
5/16	W3/0-56R		5/8	W3/0-58R
3/8	W3/0-38R		3/4	W3/0-34R
7/16	W3/0-76R			
1/2	W3/0-12R			

WIRE SIZE 4/0

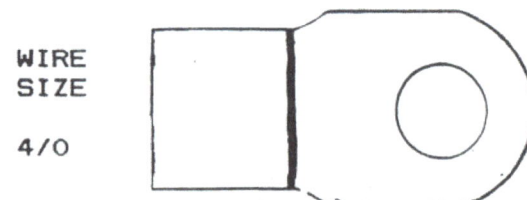 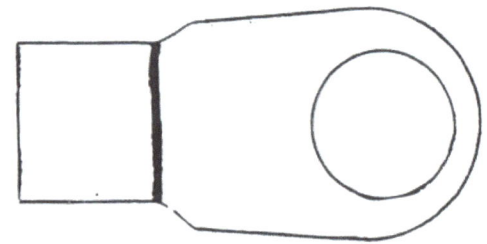

STUD	PART NO.		STUD	PART NO.
1/4	W4/0-14R		1/2	W4/0-12RW
5/16	W4/0-56R		5/8	W4/0-58R
3/8	W4/0-38R		3/4	W4/0-34R
7/16	W4/0-76R			
1/2	W4/0-12R			

MULTI-STUD

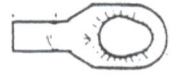

WIRE RANGE	22-18		16-14		12-10	
	STUD	PART NO.	STUD	PART NO.	STUD	PART NO.
	2-4-6	W18-246	2-4-6	W14-246	6-8-10	W10-681
	6-8-10	W18-681	6-8-10	W14-681		

Modern Electrical Devices

NON-INSULATED WITH BRAZED SEAM

WIRE RANGE 22-18

	SPADE	BLOCK SPADE	FLANGED BLOCK SPADE	HOOKS
STUD	PART NO.	PART NO.	PART NO.	PART NO.
6	W18-6S	W18-6SN	W18-6F	W18-6H
8	W18-8S	W18-8SN	W18-8F	W18-8H
10	W18-10S	W18-10SN	W18-10F	W18-10H

WIRE RANGE 16-14

	SPADE	BLOCK SPADE	FLANGED BLOCK SPADE	HOOKS
STUD	PART NO.	PART NO.	PART NO.	PART NO.
6	W14-6S	W14-6SN	W14-6F	W14-6H
8	W14-8S	W14-8SN	W14-8F	W14-8H
10	W14-10S	W14-10SN	W14-10F	W14-10H

WIRE RANGE 12-10

	SPADE	BLOCK SPADE	FLANGED BLOCK SPADE	HOOKS
STUD	STOCK #	STOCK #	STOCK #	STOCK #
6	W10-6S	W10-6SN	W10-6F	W10-6H
8	W10-8S	W10-8SN	W10-8F	W10-8H
10	W10-10S	W10-10SN	W10-10F	W10-10H
1/4	W10-14S			W10-14H
5/16	W10-56S			W10-56H

HEAVY DUTY

WIRE RANGE 16-12

STUD	STOCK #	STUD	STOCK #	STUD	STOCK #
6	W12-6S	8	W12-8S	10	W12-10S

Modern Electrical Devices

NON-INSULATED SEAMLESS

PARALLEL CONNECTORS

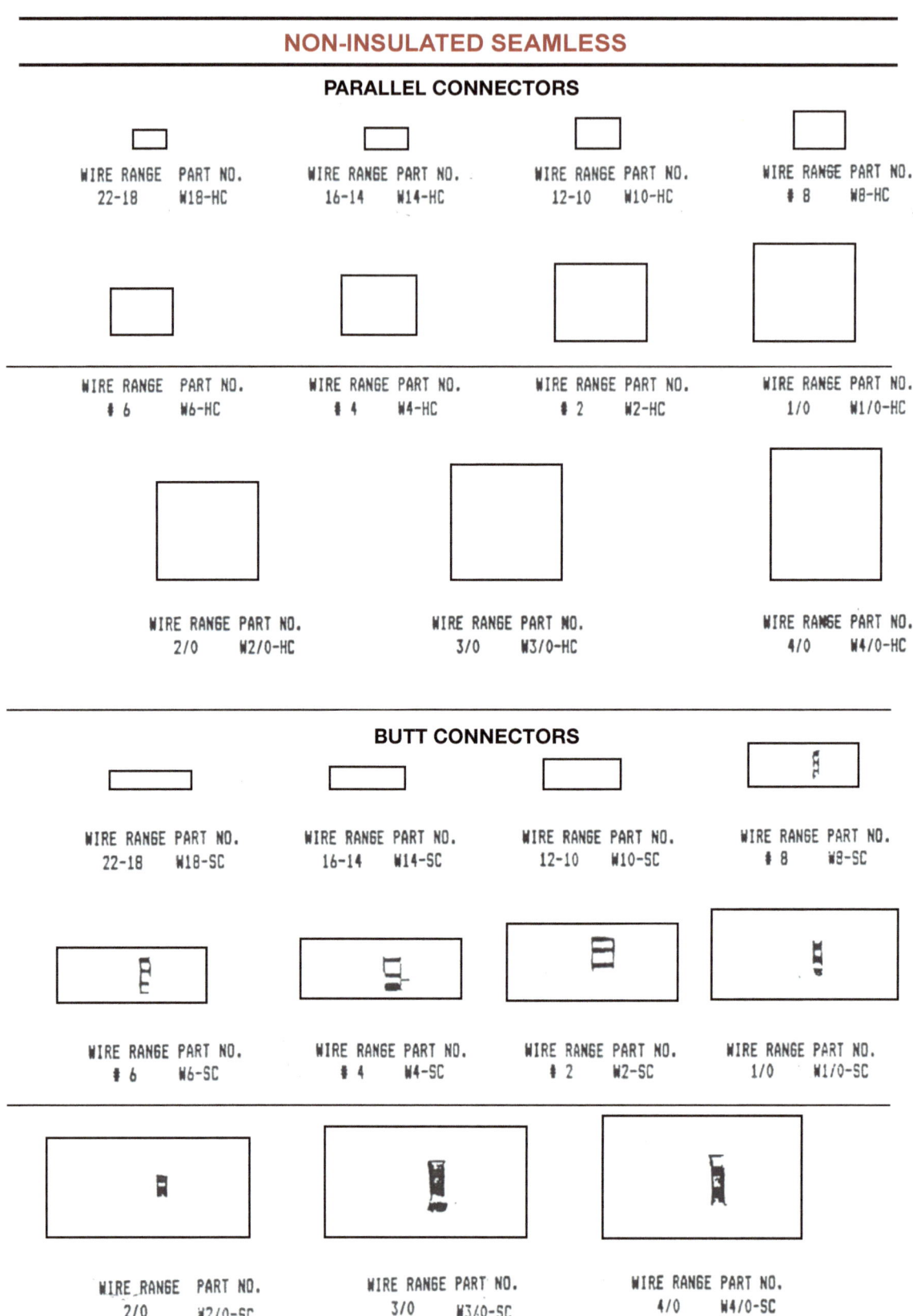

WIRE RANGE	PART NO.	WIRE RANGE	PART NO.	WIRE RANGE	PART NO.	WIRE RANGE	PART NO.
22-18	W18-HC	16-14	W14-HC	12-10	W10-HC	# 8	W8-HC
# 6	W6-HC	# 4	W4-HC	# 2	W2-HC	1/0	W1/0-HC
2/0	W2/0-HC	3/0	W3/0-HC			4/0	W4/0-HC

BUTT CONNECTORS

WIRE RANGE	PART NO.	WIRE RANGE	PART NO.	WIRE RANGE	PART NO.	WIRE RANGE	PART NO.
22-18	W18-SC	16-14	W14-SC	12-10	W10-SC	# 8	W8-SC
# 6	W6-SC	# 4	W4-SC	# 2	W2-SC	1/0	W1/0-SC
2/0	W2/0-SC	3/0	W3/0-SC			4/0	W4/0-SC

Modern Electrical Devices

NON-INSULATED WITH BUTTED SEAM

RING TONGUE TERMINALS

WIRE RANGE 22-18

STUD	PART NO.	STUD	PART NO.	STUD	PART NO.	STUD	PART NO.	STUD	PART NO.
2	E18-2RSS	2	E18-2RS	6	E18-6R	1/4	E18-14R	5/16	E18-56RW
4	E18-4RSS	4	E18-4RS	8	E18-8R	5/16	E18-56R	3/8	E18-38R
6	E18-6RSS	6	E18-6RS	10	E18-10R				

WIRE RANGE 16-14

STUD	PART NO.	STUD	PART NO.	STUD	PART NO.	STUD	PART NO.
2	E14-2RS	6	E14-6R	1/4	E14-14R	5/16	E14-56R
4	E14-4RS	8	E14-8R	5/16	E14-56RS	3/8	E14-38R
6	E14-6RS	10	E14-10R				

WIRE RANGE 12-10

STUD	PART NO.	STUD	PART NO.	STUD	PART NO.	STUD	PART NO.	STUD	PART NO.
4	E10-4RS	6	E10-6R	1/4	E10-14R	5/16	E10-56R	7/16	E10-76R
6	E10-6RS	8	E10-8R	5/16	E10-56RS	3/8	E10-38R	1/2	E10-12R
		10	E10-10R						

HEAVY DUTY RING TONGUE

WIRE RANGE 16-12

STUD	PART NO.	STUD	PART NO.	STUD	PART NO.	STUD	PART NO.	STUD	PART NO.
4	E12-4RS	6	E12-6R	1/4	E12-14RS	1/4	E12-14R	7/16	E12-76R
6	E12-6RS	8	E12-8R	5/16	E12-56RS	5/16	E12-56R	1/2	E12-12R
		10	E12-10R			3/8	E12-38R		

Modern Electrical Devices

NON-INSULATED WITH BUTTED SEAM

WIRE RANGE 22-18

SPADE	BLOCK SPADE	FLANGED BLOCK SPADE	HOOKS
STUD / PART NO.	STUD / PART NO.	STUD / PART NO.	STUD / PART NO.
6 / E18-6S	6 / E18-6SN	6 / E18-6F	6 / E18-6H
8 / E18-8S	8 / E18-8SN	8 / E18-8F	8 / E18-8H
10 / E18-10S	10 / E18-10SN	10 / E18-10F	10 / E18-10H

WIRE RANGE 16-14

SPADE	BLOCK SPADE	FLANGED BLOCK SPADE	HOOKS
STUD / PART NO.	STUD / PART NO.	STUD / PART NO.	STUD / PART NO.
6 / E14-6S	6 / E14-6SN	6 / E14-6F	6 / E14-6H
8 / E14-8S	8 / E14-8SN	8 / E14-8F	8 / E14-8H
10 / E14-10S	10 / E14-10SN	10 / E14-10F	10 / E14-10H

WIRE RANGE 12-10

SPADE	BLOCK SPADE	FLANGED BLOCK SPADE	HOOKS
STUD / PART NO.	STUD / PART NO.	STUD / PART NO.	STUD / PART NO.
6 / E10-6S	6 / E10-6SN	6 / E10-6F	6 / E10-6H
8 / E10-8S	8 / E10-8SN	8 / E10-8F	8 / E10-8H
10 / E10-10S	10 / E10-10SN	10 / E10-10F	10 / E10-10H
1/4 / E10-14S			1/4 / E10-14H
5/16 / E10-56S			5/16 / E10-56H

HEAVY DUTY

WIRE RANGE 16-12

STUD / PART NO.	STUD / PART NO.	STUD / PART NO.
6 / E12-6S	8 / E12-8S	10 / E12-10S

Modern Electrical Devices

NON-INSULATED WITH BUTTED SEAM

FEMALE QUICK CONNECTS

WIDTH 110

WIRE RANGE	PART NO.	WIRE RANGE	PART NO.	WIRE RANGE	PART NO.
26-24	E2020-110F	22-18	E1820-110F	16-14	E1420-110F
26-24	E2032-110F	22-18	E1832-110F	16-14	E1432-110F

WIDTH 187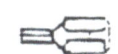

WIRE RANGE	PART NO.	WIRE RANGE	PART NO.	WIRE RANGE	PART NO.	WIRE RANGE	PART NO.
26-24	E2020-187F	22-18	E1820-187F	16-14	E1420-187F	14-12	E1220-187F
26-24	E2032-187F	22-18	E1832-187F	16-14	E1432-187F	14-12	E1232-187F

WIDTH 250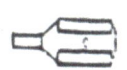

WIRE RANGE	PART NO.	WIRE RANGE	PART NO.	WIRE RANGE	PART NO.	WIRE RANGE	PART NO.
22-18	E1832-250F	16-14	E1432-250F	14-12	E1232-250F	12-10	E1032-250F

MALE QUICK CONNECTS

WIDTH 187

WIRE RANGE	PART NO.	WIRE RANGE	PART NO.
22-18	E1820-187M	16-14	E1420-187M
22-18	E1832-187M	16-14	E1432-187M

WIDTH 250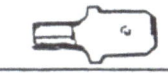

WIRE RANGE	PART NO.	WIRE RANGE	PART NO.	WIRE RANGE	PART NO.
22-18	E1832-250M	16-14	E1432-250M	12-10	E1032-250

MALE SNAP PLUG

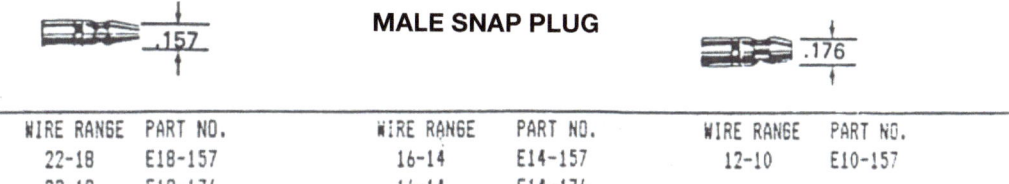

WIRE RANGE	PART NO.	WIRE RANGE	PART NO.	WIRE RANGE	PART NO.
22-18	E18-157	16-14	E14-157	12-10	E10-157
22-18	E18-176	16-14	E14-176		

Modern Electrical Devices

NON-INSULATED WITH BUTTED SEAM

MULTI STACK (PIGGYBACKS)

WIDTH 250

WIRE RANGE	PART NO.	WIRE RANGE	PART NO.
22-18	E1832-PB	16-14	E1432-PB

FLAG DISCONNECTS

WIDTH 187

WIRE RANGE	PART NO.	WIRE RANGE	PART NO.
22-18	E1820-187FL	16-14	E1420-187FL

WIDTH 250

WIRE RANGE	PART NO.	WIRE RANGE	PART NO.	WIRE RANGE	PART NO.
16-14	E1832-250FL	16-14	E1432-250FL	12-10	E1032-250FL

FLAG TERMINALS

RING — WIRE RANGE 22-18

STUD	PART NO.
6	E18-6FLR
8	E18-8FLR
10	E18-10FLR

SPADE — WIRE RANGE 22-18

STUD	PART NO.
6	E18-6FLS
8	E18-8FLS
10	E10-10FLS

RING — WIRE RANGE 16-14

STUD	PART NO.
6	E14-6FLR
8	E14-8FLR
10	E14-10FLR

SPADE — WIRE RANGE 16-14

STUD	PART NO.
6	E14-6FLS
8	E14-8FLS
10	E14-10FLS

RING — WIRE RANGE 12-10

STUD	PART NO.
6	E10-6FLR
8	E10-8FLR
10	E10-10FLR

SPADE — WIRE RANGE 12-10

STUD	PART NO.
6	E10-6FLS
8	E10-8FLS
10	E10-10FLS

Modern Electrical Devices

HIGH TEMPERATURE TERMINALS 900° NICKEL PLATED STEEL

RING TONGUE TERMINALS

WIRE RANGE 22-18

STUD	PART NO.	STUD	PART NO.	STUD	PART NO.	STUD	PART NO.
6	HT 18-6R	10	HT18-10R	1/4	HT18-14R	3/8	HT18-38R
8	HT 18-8R			5/16	HT18-56R		

WIRE RANGE 16-14

STUD	PART NO.	STUD	PART NO.	STUD	PART NO.	STUD	PART NO.
6	HT14-6RS	6	HT14-6R	1/4	HT14-14R	3/8	HT18-38R
		8	HT14-8R	5/16	HT14-56		
		10	HT14-10R				

WIRE RANGE 12-10

STUD	PART NO.	STUD	PART NO.	STUD	PART NO.	STUD	PART NO.
6	HT10-6R	10	HT10-10R	1/4	HT10-14R	3/8	HT10-38R
8	HT10-8R			5/16	HT10-56R		

SPADE TERMINAL

WIRE RANGE	STUD	PART NO.	WIRE RANGE	STUD	PART NO.	WIRE RANGE	STUD	PART NO.
22-18	10	HT18-10S	16-14	10	HT14-10S	12-10		HT10-10S

BUTT CONNECTOR

WIRE RANGE	PART NO.	WIRE RANGE	PART NO.	WIRE RANGE	PART NO.
22-18	HT18-SC	16-14	HT14-SC	12-10	HT10-SC

MALE QUICK-SLIDE

WIDTH 250

WIRE RANGE	PART NO.	WIRE RANGE	PART NO.
22-18	HT1832-250M	16-14	HT1432-250M

FEMALE QUICK-SLIDE

WIDTH 250

WIRE RANGE	PART NO.	WIRE RANGE	PART NO.	WIRE RANGE	PART NO.
22-18	HT1832-250F	16-14	HT1420-187F	16-14	HT1432-250F

Modern Electrical Devices

ALL WEATHER, HEAT-SEALABLE NYLON TERMINAL

RING TONGUE TERMINAL

WIRE RANGE	COLOR	STUD	PART NO.	WIRE RANGE	COLOR	STUD	PART NO.	WIRE RNAGE	COLOR	STUD	PART NO.
22-18	RED	6	N18-6HR	16-14	BLUE	8	N14-8HR	12-10	YEL.	10	N10-10HR
22-18	RED	8	N18-8HR	16-14	BLUE	10	N14-10HR	12-10	YEL.	1/4	N10-14HR
22-18	RED	10	N18-10HR	16-14	BLUE	1/4	N14-14HR	12-10	YEL.	5/16	N10-56HR
				16-14	BLUE	5/16	N14-56HR	12-10	YEL.	3/8	N10-38HR

SPADE TERMINAL

  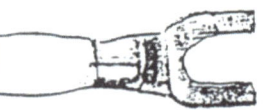

WIRE RANGE	COLOR	STUD	PART NO.	WIRE RANGE	COLOR	STUD	PART NO.	WIRE RNAGE	COLOR	STUD	PART NO.
22-18	RED	6	N18-6HS	16-14	BLUE	8	N14-8HS	12-10	YEL.	10	N10-10HS
22-18	RED	8	N18-8HS	16-14	BLUE	10	N14-10HS	12-10	YEL.	1/4	N10-14HS
22-18	RED	10	N18-10HS	16-14	BLUE	1/4	N14-14HS	12-10	YEL.	5/16	N10-56HS

BUTT CONNECTOR

WIRE RANGE	COLOR	PART NO.	WIRE RANGE	COLOR	PART NO.	WIRE RNAGE	COLOR	PART NO.
22-18	RED	N18-HSC	16-14	BLUE	N14-HSC	12-10	YEL.	N10-HSC

QUICK DISCONNECTS

WIDTH 250

WIRE RANGE	COLOR	PART NO.	WIRE RANGE	COLOR	PART NO.	WIRE RANGE	COLOR	PART NO.
22-18	RED	N1832-250HF	16-14	BLUE	N1432-250HF	12-10	YEL.	N1032-250HF

QUICK DISCONNECT COUPLERS

FEMALE   **MALE**

WIRE RANGE	DIA.	COLOR	PART NO.	WIRE RANGE	DIA.	COLOR	PART NO.
22-18	187	RED	NF1832-187HF	22-18	250	RED	NF923-250HM
16-14	187	BLUE	NF1432-187HF	16-14	250	BLUE	NF432-250HM
12-10	187	YEL.	NF1032-187HF	12-10	250	YEL.	NF032-250HM
22-18	250	RED	NF1832-250HF				
16-14	250	BLUE	NF1432-250HF				
12-10	250	YEL.	NF1032-250HF				

MALE SNAP PLUGS

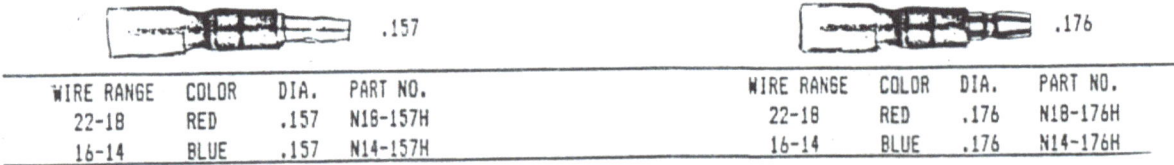

WIRE RANGE	COLOR	DIA.	PART NO.	WIRE RANGE	COLOR	DIA.	PART NO.
22-18	RED	.157	N18-157H	22-18	RED	.176	N18-176H
16-14	BLUE	.157	N14-157H	16-14	BLUE	.176	N14-176H

**MOST TERMINALS ARE AVAILABLE WITH SHRINK TUBING
CONTACT OFFICE FOR PART NUMBERS AND PRICES**

Modern Electrical Devices

SPECIAL TERMINALS

SELF STRIPPING CONNECTORS

22-18 GA. RED NO. 905	18-14 GA. BLUE NO. 801	18-16-14 GA. TO 12-10 GA. BROWN NO. 902	FOR 12-10 GA. ONLY YELLOW NO. 903	18-14 GA. BLACK NO. 901

T-TAP DISCONNECT	FUSE HOLDER	MALE FUSE CLIP ADAPTER
 PART NO. C18-250 22-18 AWG. C14-250 16-14 AWG.	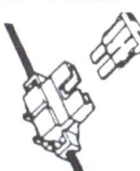 PART NO. 14 AWG. 20 AMP MAX 16 AWG. 10 AMP MAX	 PART NO. FC-250 MATES WITH MALE RECEPTACLE WIDTH .250 GAPPING .030
DOUBLE MALE-FEMALE CHAIR ADAPTER	**DOUBLE MALE-FEMALE ADAPTER**	**"SLANT" DISCONNECT**
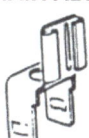 PART NO. FMA-P 2 EA. .250 MALE TABS 1 EA. .250 FEMALE TAB	 PART NO 7MMF-250 1 EA. .250 MALE TAB 1 EA. .250 FEMALE RECPTICAL	 PART NO. 77-250 2 EA. .250 MALE TABS 1 EA. .250 FEMALE TAB
PLASTIC INSULATED DOUBLE MALE QUICK-SLIDE	**OPEN BARREL MALE SPLIT TAB**	**OPEN BARREL FEMALE SPLIT TAB MALE**
 PART NO. C250-FR DOUBLE MALE RECEPTACLE .250 TAB - 2.125 LENGTH	 PART NO. 600-B WIRE RANGE 18-14 .250 TAB	 PART NO. 900-B WIRE RANGE 18-14 .250 WIDTH

WIRE NUTS			WING NUTS		
PART NO.	COLOR		PART NO.	COLOR	
329XTP-C	GRAY	71B	733TT-C	YELLOW	84
330XTP-C	BLACK	72B	735TT-C	RED	86
331XTP-C	ORANGE	73B	739TT-C	BLUE	89
333XTP-C	YELLOW	74B			
335XTP-C	RED	76B			

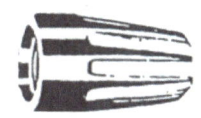
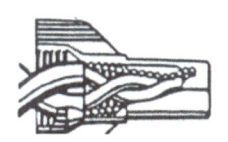

Modern Electrical Devices

STANDARD BARREL HEAVY DUTY, COPPER COMPRESSION LUGS

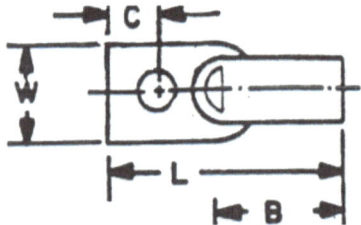

1. ONE PIECE, SEAMLESS, HIGH CONDUCTIVITY COPPER CONSTRUCTION.
2. THICK WALLED BARREL PROVIDES LOW ELECTRICAL RESISTANCE
3. PEEP HOLE DESIGN PERMITS QUICK VISUAL INSPECTION FOR PROPER INSERTION OF CABLE.
4. HOT TIN DIPPED FOR CORROSION RESISTANCE.

PART NO.	CABLE SIZE	STUD SIZE	B	W	L	C	PART NO.	CABLE SIZE	STUD SIZE	B	W	L	C
T8-10R		10	.47	.41	1.15	.21	T4/0-14R		1/4	1.00	1.02	2.69	.50
T8-14R		1/4	.47	.46	1.22	.24	T4/0-56R		5/16	1.00	1.02	2.69	.53
T8-56R	#8	5/16	.47	.57	1.30	.29	T4/0-38R	#4/0	3/8	1.00	1.02	2.44	.41
T8-38R		3/8	.47	.57	1.30	.29	T4/0-12R		1/2	1.00	1.00	2.69	.53
T8-12R		1/2	.47	.73	1.52	.39	T4/0-58R		5/8	1.00	1.00	2.83	.66
							T4/0-34R		3/4	1.00	1.00	3.14	.78
T6-10R		10	.81	.41	1.54	.22							
T6-14R	#6	1/4	.81	.41	1.57	.25	T250-14R		1/4	1.06	1.09	2.80	.56
T6-56R		5/16	.81	.52	1.79	.34	T250-56R		5/16	1.06	1.09	2.80	.56
T6-38R		3/8	.81	.62	1.84	.38	T250-38R	250 MCM	3/8	1.06	1.09	2.79	.53
							T250-12R		1/2	1.06	1.11	2.79	.53
T4-10R		#8-10	.81	.50	1.58	.25	T250-78R		7/8	1.06	1.19	3.48	.88
T4-14R		1/4	.81	.50	1.73	.31							
T4-56R	#4	5/16	.81	.58	1.92	.41	T300-56R		5/16	1.06	1.20	2.84	.56
T4-38R		3/8	.81	.58	1.92	.41	T300-38R		3/8	1.06	1.20	2.84	.53
T4-12R		1/2	.81	.75	2.12	.53	T300-12R	300 MCM	1/2	1.06	1.20	2.84	.53
							T300-58R		5/8	1.06	1.20	3.09	.66
T2-14R		1/4	.88	.61	1.88	.31	T300-78R		7/8	1.06	1.20	3.53	.88
T2-56R	#2	5/16	.88	.59	1.94	.34							
T2-38R		3/8	.88	.61	2.06	.41	T350-12R		3/8	1.12	1.29	2.95	.53
T2-12R		1/2	.88	.81	2.31	.56	T350-12R	350 MCM	1/2	1.12	1.29	2.94	.53
							T350-58R		5/8	1.12	1.29	3.20	.66
T1-14R		1/4	.88	.66	1.90	31	T350-78R		7/8	1.12	1.29	4.19	1.19
T1-56R	#1	5/16	.88	.68	1.98	.34							
T1-38R		3/8	.88	.68	1.98	.34	T400-38R		3/8	1.19	1.40	3.33	.66
T1-12R		1/2	.88	.81	2.42	.56	T400-12R	400 MCM	1/2	1.19	1.40	3.33	.66
							T400-58R		5/8	1.19	1.40	3.33	.66
T1/0-14R		1/4	.86	.75	2.03	.34	T400-78R		7/8	1.19	1.40	3.76	.88
T1/0-56R	1/0	5/16	.88	.75	2.03	.34							
T1/0-38R		3/8	.88	.75	2.16	.41	T500-38R		3/8	1.38	1.52	2.59	.66
T1/0-12R		1/2	.88	.75	2.40	.53	T500-12R		1/2	1.38	1.52	2.59	.66
							T500-58R	500 MCM	5/8	1.38	1.52	2.59	.66
T2/0-14R		1/4	.94	.83	2.27	.41	T500-34R		3/4	1.38	1.52	3.78	.75
T2/0-56R		5/16	.94	.83	2.27	.41	T500-78R		7/8	1.38	1.52	4.02	.88
T2/0-38R	#2/0	3/8	.94	.83	2.27	.41	T500-1R		1	1.38	1.52	4.20	.94
T2/0-12R		1/2	.94	.83	2.83	.69							
T2/0-58R		5/8	.94	.83	2.27	.41	T600-58R	600 MCM	5/8	1.69	1.69	4.12	.88
T2/0-34R		3/4	.94	.83	2.27	.41	T600-78R		7/8	1.69	1.69	4.17	.88
T3/0-14R		1/2	1.00	.91	2.62	.58	T750-58R	750 MCM	5/8	1.88	1.89	4.59	.88
T3/0-56R	#3/0	5/16	1.00	.91	2.31	.50	T750-78R		7/8	1.88	1.89	4.59	.88
T3/0-38R		3/8	1.00	.91	2.38	.41							
T3/0-12R		1/2	1.00	.91	2.62	.53	T1000-58R	1000 MCM	5/8	1.88	2.17	4.88	.94

Modern Electrical Devices

LONG BARREL HEAVY DUTY, COPPER COMPRESSION LUGS

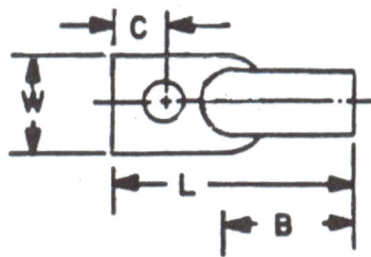

1. ONE PIECE, SEAMLESS, HIGH CONDUCTIVITY COPPER CONSTRUCTION.
2. THICK WALLED BARREL PROVIDES LOW ELECTRICAL RESISTANCE.
3. LONG BARREL INDENTATION FOR GREATER CONTACT AREA AND GREATER PULLOUT VALUES.
4. HOT TIN DIPPED FOR CORROSION RESISTANCE.

PART NO.	CABLE SIZE	STUD SIZE	B	W	L	C	PART NO.	CABLE SIZE	STUD SIZE	B	W	L	C
TL6-14R	#6	1/4	1.12	.41	1.81	.25	TL250-12R	250 MCM	1/2	1.62	1.09	3.19	.56
TL4-56R	#4	5/16	1.12	.50	1.81	.25	TL300-12R	300 MCM	1/2	2.00	1.19	3.56	.56
TL2-56R	#2	5/16	1.25	.59	2.13	.25	TL350-12R	350 MCM	1/2	2.00	1.28	3.72	.56
TL1-56R	#1	5/16	1.38	.67	2.38	.38	TL400-58R	400 MCM	5/8	2.12	1.38	4.25	.75
TL1/0-56R	1/0	5/16	1.38	.75	2.38	.38	TL500-58R	500 MCM	5/8	2.25	1.52	4.38	.75
TL2/0-38R	2/0	3/8	1.50	.81	2.68	.44	TL600-58R	600 MCM	5/8	2.69	1.69	5.25	.88
TL3/0-12R	3/0	1/2	1.50	.91	2.84	.50	TL750-58R	750 MCM	5/8	2.88	1.89	5.59	.88
TL4/0-12R	4/0	1/2	1.62	1.00	2.97	.50	TL1000-58R	1000 MCM	5/8	3.00	2.17	6.00	.94

STANDARD BARREL COMPRESSION CONNECTORS

LONG BARREL DOUBLE INDENT COMPRESSION CONNECTORS

PART NO.	CABLE SIZE	L	B		PART NO.	CABLE SIZE	L	B
T8-SC	#8	1.06	.47		TL6-SC	#6	2.38	1.12
T6-SC	#6	1.75	.81		TL4-SC	#4	2.38	1.12
T4-SC	#4	1.75	.81		TL2-SC	#2	2.62	1.25
T2-SC	#2	1.88	.88		TL1-SC	#1	2.88	1.38
T1-SC	#1	1.88	.88					
T1/0-SC	1/0	1.88	.88		TL1/0-SC	1/0	2.88	1.38
T2/0-SC	2/0	2.00	.94		TL2/0-SC	2/0	3.12	1.50
T3/0-SC	3/0	2.12	1.00	REGULAR	TL3/0-SC	3/0	3.12	1.50
T4/0-SC	4/0	2.12	1.00		TL4/0-SC	4/0	3.38	1.62
T250-SC	250 MCM	2.25	1.06		TL250-SC	250 MCM	3.38	1.62
T300-SC	300 MCM	2.25	1.06		TL300-SC	300 MCM	4.12	2.00
T350-SC	350 MCM	2.38	1.12		TL350-SC	350 MCM	4.12	2.00
T400-SC	400 MCM	2.50	1.19		TL400-SC	400 MCM	4.38	2.12
T500-SC	500 MCM	2.88	1.38		TL500-SC	500 MCM	4.62	2.25
T600-SC	600 MCM	2.88	1.38		TL600-SC	600 MCM	5.50	2.69
T750-SC	750 MCM	3.38	1.62		TL750-SC	750 MCM	5.58	2.68
T1000-SC	1000 MCM	3.88	1.88	LONG	TL1000-SC	1000 MCM	6.12	3.00
					TL1500-SC	1500 MCM	6.50	3.19
					TL2000-SC	2000 MCM	7.00	3.53

Modern Electrical Devices

SOLDER LUGS

PART NO.	AWG WIRE SIZE	STUD SIZE "A"	LENGTH "B"	WIDTH OF PAD "C"	CLEARANCE "D"	BARREL O.D. "E"	BARREL I.D. "E"	AMP CAPACITY
ST6-10R ST6-14R ST6-56R	6	10 1/4 5/16	1-7/32	7/16	11/32	.312	.232	50
ST4-10R ST4-14R ST4-38R	4	10 1/4 5/16 3/8	1-11/32	17/32	13/32	.375	.286	70
ST2-14R ST2-56R ST2-38R	2	1/4 5/16 3/8	1-1/2	5/8	13/32	.437	.336	90
ST1-14R ST1-56R ST1-38R ST1-12R	1	1/4 5/16 3/8 1/2	1-9/16	5/8	13/32	.465	.360	100
ST1/0-14R ST1/0-56R ST1/0-38R ST1/0-12R	1/0	1/4 5/16 3/8 1/2	1-3/4	45/64	13/32	.515	.407	125
ST2/0-14R ST2/0-56R ST2/0-38R ST2/0-12R	2/0	1/4 5/16 3/8 1/2	2	13/16	1/2	.570	.461	150
ST3/0-1/4R ST3/0-56R ST3/0-38R ST3/0-12R	3/0	1/4 5/16 3/8 1/2	2-1/8	29/32	1/2	.630	.511	175
ST4/0-14R ST4/0-56R ST4/0-38R ST4/0-12R	4/0	1/4 5/16 3/8 1/2	2-9/32	1	17/32	.690	.559	225
ST250-56R ST250-38R ST250-12R	250 MCM	5/16 3/8 1/2	2-5/8	1-3/16	5/8	.815	.685	250
ST400-38R	400 MCM	3/8	3-3/8	1-13/32	7/8	.937	.776	325
ST450-38R	450 MCM	3/8	3-3/8	1-1/2	15/16	1.000	.820	362
ST500-38R	500 MCM	3/8	3-7/16	1-9/16	1-3/16	1.062	.880	400
ST600-12R	600 MCM	1/2	4-1/16	1-11/16	1-1/4	1.125	.943	-450
ST800-12R	800 MCM	1/2	5	1-15/16	1-3/8	1.312	1.084	550
ST1000-78R	1000 MCM	7/8	5-3/8	2-1/8	1-5/16	1.437	1.210	650

Modern Electrical Devices

STANDARD NYLON CABLE TIES

PACKAGED PART NO. CLEAR	PACKAGED PART NO. BLACK	PKG. QTY.	BULK PART NO. CLEAR	BULK PART NO. BLACK	BULK QTY.	MAX. BUNDLE DIAMETER	MAX. LENGTH	MAX. TENSILE STRENGTH
A18-4C	A18-4UVC	100	A18-4	A18-4UV	1000	13/16 IN.	4 IN.	18 LBS.
A18-6C	A18-6UVC	100	A18-6	A18-6UV	1000	1 3/8 IN.	8 IN.	18 LBS.
A30-6C	A30-6UVC	100	A30-6	A30-6UV	1000	1 3/8 IN.	6 IN.	40 LBS.
A50-8C	A50-8UVC	100	A50-8	A50-8UV	1000	2 IN.	8 IN.	75 LBS.
A50-12C	A50-12UVC	100	A50-12	A50-12UV	1000	3 3/16 IN.	11 IN.	75 LBS.
A50-15C	A50-15UVC	100	A50-15	A50-15UV	1000	4 IN.	14 IN.	75 LBS.
A120-8C	A120-8UVC	100	A120-8	A120-8UV	1000	2 IN.	8 1/4 IN.	175 LBS.
A120-12C	A120-12UVC	100	A120-12	A120-12UV	1000	3 1/4 IN.	12 1/4 IN.	175 LBS.
A120-15C	A120-15UVC	100	A120-15	A120-15UV	500	4 IN.	14 1/4 IN.	175 LBS.
A175-18L	A175-18UVL	50	A175-18	A175-18UV	500	5 1/4 IN.	18 1/4 IN.	175 LBS.
A175-21L	A175-21UVL	50	A175-21	A175-21UV	500	6 IN.	21 1/4 IN.	175 LBS.
A175-24L	A175-24UVL	50	A175-24	A175-24UV	500	7 IN.	24 1/4 IN.	175 LBS.
A175-34L	A175-34UVL	50	A175-34	A175-34UV	500	10 IN.	34 IN.	175 LBS.
A175-36L	A175-36UVL	50	A175-36	A175-36UV	500	10 3/4 IN.	36 IN.	175 LBS.
A175-41L	A175-41UVL	50	A175-41	A175-41UV	500	12 IN.	41 IN.	175 LBS.
A175-48L	A175-48UVL	50	A175-48	A175-48UV	500	15 IN.	48 IN.	175 LBS.

RELEASABLE TIES

A50R-9C	A50R-9UVC	100	A50R-9	A50R-9UV	1000	2 3/4 IN.	9 IN.	50 LBS.
A50R-12C	A50R-12UVC	100	A50R-12	A50R-12UV	1000	3 IN.	12 IN.	50 LBS.
A50R-15C	A50R-15UVC	100	A50R-15	A50R-15UV	1000	4 IN.	15 IN.	50 LBS.

MOUNTING NYLON CABLE TIES

A50M-9	A50M-9UVC	100	A50M-9	A50M-9UV	1000	2 3/4 IN.	9 IN.	50 LBS.
A50M-12C	A50M-12UVC	100	A50M-12	A50-12UV	1000	3 IN.	12 IN.	50 LBS.
A50M-15C	A50M-15UVC	100	A50M-15	A50-15UV	500	4 IN.	15 IN.	50 LBS.
A120M-15C	A120M-15UVC	100	A120M-15	A120M-15UV	100	4 IN.	15 IN.	120 LBS.
A175M-15L	A175M-15UVL	50	A175M-15	A175M-15UV	50	4 IN.	15 IN.	175 LBS.

STANDARD TIES AVAILABLE IN COLORS AND FLUORESCENT COLORS • COMPLETE LINE OF TOOLS FOR TENSIONS AND INSTALLING TIES AVAILABLE

NYLON ADHESIVE-BACKED MOUNTING BASES WITH SCREW HOLES

PART NO.	BASE SIZE	TIE RANGE	SCREW SIZE
A1830	3/4 IN.SQ.	A18	#6
A4050	1 1/8 IN.SQ.	A40 TO A75	#6
A1200	1 1/2 IN.SQ.	A120 TO A175	#8

Modern Electrical Devices

CABLE WRAP

MATERIAL	PART NO.	OUTSIDE DIAMETER	WALL	PITCH	MAXIMUM BUNDLE RANGE
CLEAR POLY-ERTHLENE	HT 1/8C HT 1/4C HT 3/8C HT 1/2C HT 3/4C HT 1C	.125 .250 .375 .500 .750 1.00	.032 .045 .052 .062 .065 .095	.187 .375 .438 .563 .750 1.00	1/6 to 1/2 3/16 to 2 5/16 to 3 3/8 to 4 3/4 to 5 1 to 7
BLACK POLY-ERTHLENE	HT 1/8UR HT 1/4UR HT 3/8UR HT 1/2UR HT 3/4UR HT 1UR	.125 .250 .375 .500 .750 1.00	.032 .045 .052 .062 .065 .095	.187 .375 .438 .500 .750 1.00	1/16 to 1/2 3/16 to 2 5/16 to 3 3/8 to 4 3/4 to 5 1 to 7
FIRE RESISTANT	HT 1/8FR HT 1/4FR HT 3/8FR HT 1/2FR HT 3/4FR HT 1FR	.125 .250 .375 .500 .750 1.00	.032 .045 .052 .062 .065 .095	.187 .375 .438 .563 .750 1.00	1/16 to 1/2 3/16 to 2 5/16 to 3 3/8 to 4 3/4 to 5 1 to 7
	HT 1/4FR-B HT 1/2FR-B	.250 .500	.045 .062	.375 .563	3/16 to 2 3/8 to 4

MATERIAL	PART NO.	OUTSIDE DIAMETER	WALL	PITCH	MAXIMUM BUNDLE RANGE
NYLON	HT 1/8N HT 1/4N HT 3/8N HT 1/2N HT 3/4N HT 1N	.125 .250 .375 .500 .750 1.00	.015 .025 .035 .035 .032 .032	.187 .375 .438 .500 .750 1.00	1/16 to 1/2 3/16 to 2 5/16 to 3 3/8 to 4 1/2 to 5 1 to 7
BLACK NYLON	HT 1/8N-B HT 1/4N-B HT 3/8N-B HT 1/2N-B HT 3/4N-B HT 1N-B	.125 .250 .375 .500 .750 1.00	.015 .025 .035 .035 .032 .032	.187 .375 .438 .500 .750 1.00	1/16 to 1/2 3/16 to 2 5/16 to 3 3/8 to 4 1/2 to 5 1 to 7
TEFLON	HT 1/8T HT 3/16T HT 1/4T HT 3/8T HT 1/2T HT 3/4T HT 1T	.125 .187 .250 .375 .500 .750 1.00	.020 .030 .030 .030 .030 .032 .040	.187 .250 .375 .438 .500 .750 1.00	1/16 to 1/2 1/8 to 1-1/2 3/16 to 2 5/16 to 2-1/2 3/8 to 3 1/2 to 4 3/4 to 5

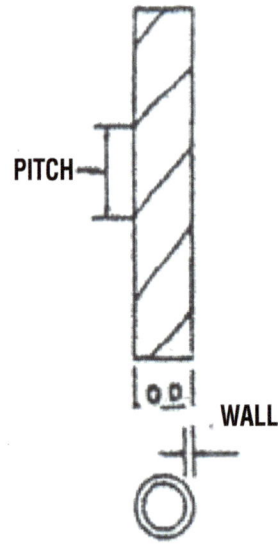

Modern Electrical Devices

IRRADIATED POLYOLEFIN SHRINKABLE TUBING

CONFORMS TO MIL-1-230538/5
SHRINKS 50% (2:1 RATIO) WITH ONLY 5% LONGITUDIBAK SGRUBJAGE
FEWER SIZES COVERING MORE APPLICATIONS
OPERATION TEMP. RANGE -55 C TO +135 C
DIELECTRIC STRENGTH: 500V PER MIL.
STOCK COLORS: BLACK, WHITE, RED, YELLOW, BLUE, AND CLEAR

PART NUMBER	SIZE (INCHES)	MIN. I.D. BEFORE SHRINKAGE	RECOVERED I.D. AFTER SHRINKAGE	RECOVERED NOM. WALL THICKNESS
221-3/64"	3/64	.046	.023	.016
221-1/16"	1/16	.063	.031	.017
221-3/32"	3/32	.093	.046	.020
221-1/8"	1/8	.125	.062	.020
221-3/16"	3/16	.187	.093	.020
221-1/4"	1/4	.250	.125	.025
221-3/8"	3/8	.375	.187	.025
221-1/2"	1/2	.500	.250	.025
221-3/4"	3/4	.750	.375	.030
221-1	1	1.000	.500	.035
221-1-1/2"	1-1/2	1.500	.750	.040
221-2	2	2.000	1.000	.045
221-3	3	3.000	1.500	.050
211-4	4	4.000	2.000	.055

POLYVINYLCHLORIDE SHRINKABLE TUBING

PART NO.	SIZE IN.	MIN. I.D. BEFORE SHRINKING IN.	RECOVERED I.D. AFTER SHRINKING	NOMINAL WALL THICKNESS
205-3/64"	3/64	.046	.023	.020
205-1/16"	1/16	.062	.031	.020
205-3/32"	3/32	.093	.046	.020
205-1/8"	1/8	.125	.062	.025
205-3/16"	3/16	.187	.093	.025
205-1/4"	1/4	.250	.125	.025
205-3/8"	3/8	.375	.187	.030
205-1/2"	1/2	.500	.250	.030
205-3/4"	3/4	.775	.375	.035
205-1"	1	1.000	.500	.040

Modern Electrical Devices

HEAT SHRINKABLE OR INSULATION CONNECTOR COVERS FOR COMPRESSION SPLICES, SELF HEALING, THICK WALL

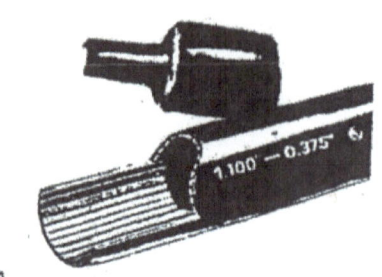

PART NUMBER	MINIMUM EXPANDED I.D.	WALL THICKNESS MINIMUM	RECIVERED I.D. MAXIMUM	RECOVERED WALL (NOMINAL)	STANDARD LENGTH (INCHES)
SPC 400	.400"	.030"	.150"	.060"	12,24
SPC 800	.800"	.025"	.220"	.085"	12,24
SPC 110	1.1"	.040"	.375"	.105"	12,24
SPC 150	1.5"	.050"	.500"	.120"	12,24
SPC 180	1.8"	.030"	.700"	.100"	12,24
SPC 200	2.0"	.050"	.750"	.140"	12,24
SPC 300	3.0"	.050"	1.120"	.155"	12,24

IRRADIATED POLYOLEFIN MULTIPLE WALL SHRINKABLE TUBING - INNER WALL MELTS

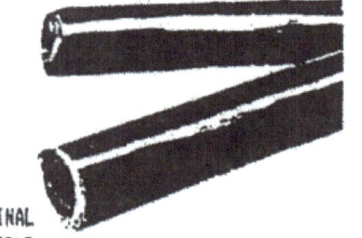

PART NUMBER	MINIMUM I.D. BEFORE SHRINKAGE	RECOVERED I.D. AFTER SHRINKAGE	RECOVERED NOMINAL WALL THICKNESS	NOMINAL MELTABLE WALL
300-1/8	.125	.023	.038	.020
300-3/16	.187	.060	.043	.025
300-1/4	.250	.080	.047	.027
300-300	.300	.050	.100	.065
300-3/8	.375	.135	.050	.030
300--1/2	.500	.195	.055	.035
300-3/4	.750	.313	.065	.040
300-1"	1.00	.400	.075	.040

Modern Electrical Devices

SPLIT BOLT CONNECTORS

"UNPLATED" TYPE - FOR COPPER TO COPPER					
PART NO.	WIRE RANGE (AWG OR MCM)				WIRE SIZES NO. 12 - 1000 MCM
	MAN		TAP		
	MAX.	MIN.	MAX.	MIN.	
TL-10	10 str.	12 sol.	10 str.	14 sol.	
TL-8	8 str.	10 sol.	8 str.	14 sol.	
TS-6	6 sol.	10 sol.	6 sol.	10 sol.	
*TSS-6	6 sol.	8 sol.	6 sol.	12 sol.	
TS-4	4 sol.	8 sol.	4 sol.	12 sol.	
*TSS-4	4 sol.	8 sol.	4 sol.	12 sol.	
TS-3	4 str.	6 sol.	4 sol.	12 sol.	
TS-2	2 sol.	6 sol.	2 sol.	10 sol.	
*TSS-2	2 sol.	4 sol.	2 sol.	10 sol.	
TS-2ST	2 str.	2 sol.	2 str.	10 sol.	
TS-1/0	1/0	2 sol.	1/0	10 sol.	
TL-2/0	2/0	2 str.	2/0	10 sol.	
TL-3/0	4/0	1/0	3/0	10 sol.	
TL-4/0	250	2/0	250	8 sol.	
TL-350	350	4/0	350	8 sol.	
TL-500	500	250	500	8 sol.	
TL-750	750	350	750	8 sol.	
TL-1000	1000	500	1000	8 sol.	

*WILL TAKE 3 WIRES OF RATED SIZE FOR USE ON R.E.A. DEAD ENDS (COPPER OR COPPERWELD)

Modern Electrical Devices

SOLDERLESS LUGS

Part No.	Wire Size Max.	Wire Size Min.	Stud Size	(CL)	Entry Dia. (B)	(g)	Height (k)	Width (H)	Length (F)	WIRE SIZES NO. 14 - 1000 MCM

PLAIN COPPER FINISH, SCREWDRIVER SLOT SCREW

Part No.	Max.	Min.	Stud Size	(CL)	Entry Dia. (B)	(g)	Height (k)	Width (H)	Length (F)	
SS-35	8 str.	14 sol.	10	5/16"	5/32	3/16	3/8	3/8	27/32	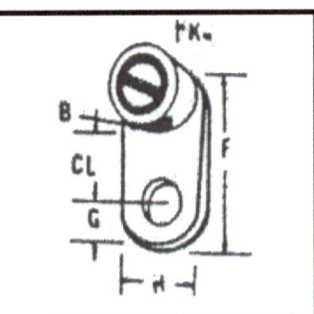
SS-70	4 str.	14 sol.	1/4	3/8"	9/32	9/32	17/32	17/32	1-3/16	

PLAIN COPPER FINISH, SOCKET HEAD SCREW

Part No.	Max.	Min.	Stud Size	(CL)	Entry Dia. (B)	(g)	Height (k)	Width (H)	Length (F)	
SS-125	1/0 str.	6 sol.	5/16	15/32	13/32	5/16	3/4	47/64	1-1/2	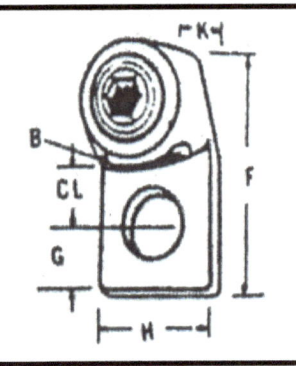
SS-250	250 MCM	1/0 str.	3/8	21/32	5/8	15/32	1-	59/64	2-	
SS-400	500 MCM	4/0 str.	1/2	1	7/8	5/8	1-3/8	1-3/8	2-15/16	
SS-650	1000 MCM	500 MCM	1/2	1-3/16	1-3/16	3/4	1-29/32	2-1/32	3-17/32	

Part No.	Wire Size Max.	Wire Size Min.	Stud Size	Clearance (CL)	(g)	Width (H)	Length (F)	WIRE SIZES NO. 14 - 500 MCM

PLAIN COPPER FINISH, SLOT HEX SCREW

Part No.	Max.	Min.	Stud Size	(CL)	(g)	Width (H)	Length (F)	
SO-8	8	14	10	9/32	7/32	3/8	1-1/16	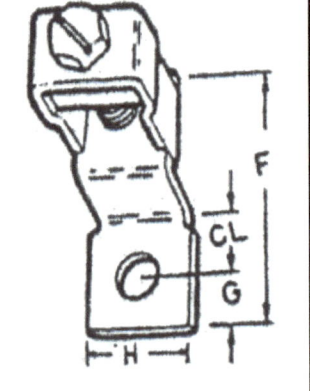
SO-4	4	14	1/4	13/32	9/32	17/32	1-13/32	

PLAIN COPPER FINISH, HEX HEAD SCREW

Part No.	Max.	Min.	Stud Size	(CL)	(g)	Width (H)	Length (F)	
SO-1/0	1/0	6	5/16	13/32	13/32	3/4	1-13/16	
SO-4/0	4/0	2	3/8	5/8	17/32	31/32	2-7/16	
SO-500	500 MCM	4/0	7/16	1-3/16	15/16	1-9/16	4-3/16	

Modern Electrical Devices

NYLON CABLE CLAMPS

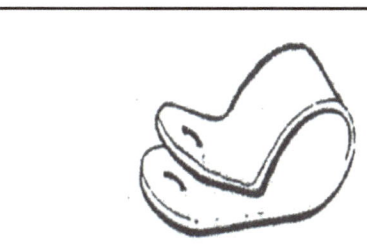

PART NO.	INSIDE DIAMETER	WIDTH	MOUNTING HOLE
N18-N	1/8"		
N36-N	3/16"		
N14-N	1/4"	3/8"	11/64"
N56-N	5/16"		
N38-N	3/8"		
N14	1/4"		
N56	5/16"		
N38	3/8"	1/2"	13/64"
N76	7/16"		
N12	1/2"		
N96	9/16"		
N58	5/8"		
N116	11/16"	1/2"	13/64"
N34	3/4"		
N78	7/8"		
N1	1"		
N1-18	1-1/8"		
N1-36	1-3/16"		
N1-14	1-1/4"	1/2"	13/64"
N1-38	1-3/8"		
N1-12	1-1/2"		
N1-34	1-3/4"		
N-2	2"		

PLASTIC INSULATED METAL CLAMPS

Zinc plated steel with a 1/32" black vinyl plastisol coating

PART NO.	INSIDE DIAMETER	WIDTH	MOUNTING HOLE
RC18-N	1/8"		
RC14-N	1/4"	1/2"	9/32"
RC56-N	5/16"		
RC38-N	3/8"		
RC56	5/16"		
RC38	3/8"		
RC76	7/16"		
RC12	1/2"	3/4"	13/32"
RC58	5/8"		
RC34	3/4"		
RC78	7/8"		
RC1	1"		
RC1-14	1-1/4"		
RC1-38	1-3/8"		
RC1-12	1-1/2"	3/4"	13/32"
RC1-58	1-5/8"		
RC1-34	1-3/4"		
RC2	2"	3/4"	13/32"

ADHESIVE BACKED CLAMPS

CAT. NO.	A (inches)
KC-18	1/8
KC-14	1/4
KC-56	5/16
KC-38	3/8
KC-12	1/2
KC-34	3/4

Modern Electrical Devices

HEAVY DUTY COPPER, CRIMP OR SOLDER BATTERY TERMINALS

STRAIGHT

PART NO.	CABLE SIZE	POST TYPE
HDBT-6U	#6	UNIVERSAL
HDBT-4U	4	UNIVERSAL
HDBT-4P	4	POSITIVE
HDBT-4N	4	NEGATIVE
HDBT-1P	1 and 2	POSITIVE
HDBT-1N	1 and 2	NEGATIVE
HDBT-1/0P	1/0	POSITIVE
HDBT-1/0N	1/0	NEGATIVE
HDBT-2/0P	2/0	POSITIVE
HDBT-2/0N	2/0	NEGATIVE
HDBT-3/0P	3/0	POSITIVE
HDBT-3/0N	3/0	NEGATIVE
HDBT-4/0P	4/0	POSITIVE
HDBT-4/0N	4/0	NEGATIVE

90° RIGHT ANGLE

PART NO.	CABLE SIZE	POST TYPE
RA-1/0P	1/0	POSITIVE
RA-1/0N	1/0	NEGATIVE
RA-2/0P	2/0	POSITIVE
RA-2/0N	2/0	NEGATIVE
RA-3/0P	3/0	POSITIVE
RA-3/0N	3/0	NEGATIVE
RA-4/0P	4/0	POSITIVE
RA-4/0N	4/0	NEGATIVE

ELBOWS RIGHT

PART NO.	CABLE SIZE	POST TYPE
EL-1/0P	1/0	POSITIVE
EL-1/0N	1/0	NEGATIVE
EL-2/0P	2/0	POSITIVE
EL-2/0N	2/0	NEGATIVE
EL-3/0P	3/0	POSITIVE
EL-3/0N	3/0	NEGATIVE
EL-4/0P	4/0	POSITIVE
EL-4/0N	4/0	NEGATIVE

ELBOWS LEFT

PART NO.	CABLE SIZE	POST TYPE
TH-4U	#4	UNIVERSAL
TH-1U	#1	UNIVERSAL
TH-1/0P	1/0	POSITIVE
TH-1/0N	1/0	NEGATIVE
TH-1/0P	1/0	POSITIVE
TH-1/0N	1/0	NEGATIVE
TH-2/0P	2/0	POSITIVE
TH-2/0N	2/0	NEGATIVE
TH-3/0P	3/0	POSITIVE
TH-3/0N	3/0	NEGATIVE
TH-4/0P	4/0	POSITIVE
TH-4/0N	4/0	NEGATIVE

STRAIGHT

RIGHT ELBOW

LEFT ELBOW

90° RIGHT ANGLE

UNIVERSAL BATTERY TERMINALS

PART NO.	GA.	PART NO.	GA.	PART NO.	GA.	PART NO.	GA.	PART NO.	DESCRIPTION
888-L	4-2/0	HD888-L	1-3/0	SM-888-L	6-1	WN-566 5/16	4-2/0	BNB	NUT & BOLT
						WN-386 3/8	4-2/0	BNB-C	SHOULDER NUT & BOLT
REGULAR		HEAVY DUTY		SIDE MOUNT W/BOLT		WING NUT		BATTERY NUT & BOLT	

www.ingramcontent.com/pod-product-compliance
Lightning Source LLC
Chambersburg PA
CBHW050838180526
45159CB00004B/1952